RHABDOVIRUSES

Volume I

Editor

David H. L. Bishop
Professor
Department of Microbiology
University of Alabama in Birmingham
Birmingham, Alabama

CRC Press
Taylor & Francis Group
Boca Raton London New York

CRC Press is an imprint of the
Taylor & Francis Group, an **informa** business

First published 1979 by CRC Press
Taylor & Francis Group
6000 Broken Sound Parkway NW, Suite 300
Boca Raton, FL 33487-2742

Reissued 2018 by CRC Press

Library of Congress Cataloging in Publication Data
Main entry under title:

Rhabdoviruses.

Includes bibliographies and indexes.
1. Rhabdoviruses. I. Bishop, David H. L. [DNLM:
1. Rhabdoviruses. QW168.5.R4 R468]
QR415.R46 576'.6484 79-20575
ISBN 0-8493-5913-9 (v. 1)

A Library of Congress record exists under LC control number: 79020575

Publisher's Note
The publisher has gone to great lengths to ensure the quality of this reprint but points out that some imperfections in the original copies may be apparent.

Disclaimer
The publisher has made every effort to trace copyright holders and welcomes correspondence from those they have been unable to contact.

ISBN 13: 978-1-315-89728-8 (hbk)
ISBN 13: 978-1-351-07638-8 (ebk)

Visit the Taylor & Francis Web site at http://www.taylorandfrancis.com and the
CRC Press Web site at http://www.crcpress.com

FOREWORD

The Rhabdoviridae is a diverse family of enveloped RNA viruses (comprising 70 or more known serotypes). Member viruses infect homeothermic or poikilothermic vertebrates, invertebrates, or plants. Particular viruses are discussed in Volumes I and III in this series. Although some of the viruses are antigenically related to each other (so far only three genera of rhabdoviruses have been recognized), for many, no antigenic cross reactivities can be detected using the techniques commonly employed to categorize viruses.

A striking feature of the Rhabdoviridae family is the diversity of the host species that are susceptible to virus infection. Some rhabdoviruses are arboviruses while others have no arthropod host or vector (Volume I). Certain members of the Rhabdoviridae family cause severe diseases of man (e.g. rabies, see Volume I), domestic animals (e.g. bovine ephemeral fever virus, Volume III), fish (e.g. spring viremia of carp virus, Volume III), or plants (Volume III). Many other rhabdoviruses appear to be innocuous in their primary host (e.g. sigma virus, Volume III).

Due to their economic and disease importance, a considerable amount of research into the molecular biology of rhabdoviruses has been undertaken in many countries of the world over the past decade. While there are several important gaps in our understanding, what is known about virus structure (Volume I), the viral infection processes, assay systems, growth potential, molecular biology, and genetics (Volume II) of representative viruses are described in this series.

The process of interference of rhabdovirus replication by defective interfering virus particles has been a feature of rhabdoviruses which has received significant attention over the last few years. The subject is still under investigation and although the developments of the last few months are not covered in this series, background information on defective particle generation, replication and interference capabilities are discussed in Volume II. Other features of rhabdovirus infections which are discussed in this series are their ability to kill cells, form pseudotypes, and establish persistent infections (Volume III).

No book series on rhabdoviruses would be complete without a discussion of virus vaccines and other possible therapeutic processes. The advancements made in rabies vaccine development over the past years are described in Volume III. The development of vaccines for other rhabdoviruses will presumably be governed by the importance of the diseases they cause.

The goal of this book series has been to provide an overview of rhabdovirology as a whole (including an appraisal of current research findings), suitable for students, teachers, and research workers. To realize this goal I asked many of the research leaders in the different disciplines of rhabdovirology to contribute chapters. Only a few were not able to particpate due to prior commitments; most of those asked responded with articles which I believe do justice to what is known about the subject.

A final point, the four International Rhabdovirus Symposia that have been held every two years since 1973 have brought together rhabdovirologists from different parts of the world and from different disciplines of the field; the Symposia have been a major factor in the development of the subject and the communication that exists among the research scientists. This book series is therefore dedicated to those who had the foresight to initiate these Symposia, the participants of the meetings, and to the contributors who have given their time and energies to the compilation of this work.

David H. L. Bishop
October 1979

THE EDITOR

David H. L. Bishop, Ph.D., is Professor of Microbiology in the Medical Center of the University of Alabama in Birmingham, Alabama. He is also a Senior Scientist in the Medical School Comprehensive Cancer Center and Diabetes Research and Training Center.

Dr. Bishop was graduated from the University of Liverpool, England, with a B.Sc. (Hons.) degree in Biochemistry in 1959. He received a Ph.D. (Biochemistry) in 1962 also from the University of Liverpool. After a postdoctoral year sponsored by a Research Fellowship at the Centre Nationale de la Recherche Scientifique, Gif-sur-Yvette, France, Dr. Bishop was a Research Associate from 1963 to 1966 in the Department of Zoology, Edinburgh University, a Research Fellow from 1966 to 1969 in the Department of Microbiology, University of Illinois, Assistant Professor (1969 to 1970) then Associate Professor (1970 to 1971) at Columbia University College of Physicians and Surgeons, Department of Human Genetics and Development. Before joining the faculty as a Professor of Microbiology at the Medical Center of the University of Alabama in Birmingham in the fall of 1975, Dr. Bishop was an Associate Professor (1971 to 1975), then Professor (1975) at the Waksman Institue of Microbiology, Rutgers University.

Dr. Bishop has published more than 100 research papers over his career and has been on the Editorial Board of the *Journal of Virology,* since 1974, and *Virology* since 1979. In addition to being active in the American Society of Microbiology and the American Society of Tropical Medicine and Hygiene, Dr. Bishop is Chairman of the Bunyaviridae Study Group of the International Committee for Taxonomy of Viruses as well as a member of its Rhabdoviridae Study Group. He has also been a member of the National Cancer Institute Scientific Review Committee from 1975 to 1979.

CONTRIBUTORS

David H. L. Bishop
Professor
Department of Microbiology
University of Alabama in Birmingham
Birmingham, Alabama

Fred Brown
Head, Department of Biochemistry
Animal Virus Research Institute
Pirbright, Surrey, England

H Fred Clark
Associate Professor
Wistar Institute
Philadelphia, Pennsylvania

J. P. Clewley
Research Fellow
Department of Biological Sciences
University of Warwick
Coventry, England

Philip Cohen
Reader in Biochemistry
University of Dundee
Dundee, Scotland

Joan Crick
Principal Scientific Officer
Animal Virus Research Institute
Pirbright, Surrey, England

Suzanne U. Emerson
Associate Professor
Microbiology Department
University of Virginia School of Medicine
Charlottesville, Virginia

James R. Etchison
Research Associate
Department of Cell, Viral, and Molecular Biology
University of Utah Medical Center
Salt Lake City, Utah

Christina L. Frazier
Assistant Professor of Biology
Southeast Missouri State University
Cape Girardeau, Missouri

Alyne K. Harrison
Research Biologist — Electron Microscopist
Viral Pathology Branch
Center for Disease Control
Atlanta, Georgia

John T. Heyward
Research Microbiologist
Viral Exanthems Branch (7-240)
Virology Division, Bureau of Laboratories
Center for Disease Control
Atlanta, Georgia

Brian P. Holloway
Microbiologist, Research Technician
Viral Exanthems Branch
Center for Disease Control
Atlanta, Georgia

D. M. Hunt
Research Fellow
Department of Biochemistry
The University of Mississippi Medical Center
Jackson, Mississippi

James J. McSharry
Associate Professor
Department of Microbiology/Immunology
Albany Medical College
Albany, New York

Margaret D. G. Mellon
Research Associate
Microbiology Department
University of Virginia School of Medicine
Charlottesville, Virginia

Frederick A. Murphy
Associate Dean
College of Veterinary Medicine and Biomedical Sciences
Colorado State University
Fort Collins, Colorado

CONTRIBUTORS (continued)

John F. Obijeski
Research Microbiologist
Viral Exanthems Branch (7-240)
Virology Division, Bureau of Laboratories
Center for Disease Control
Atlanta, Georgia

Robert E. Shope
Professor of Epidemiology
Yale Arbovirus Research Unit
Department of Epidemiology and Public Health
Yale School of Medicine
New Haven, Connecticut

Donald F. Summers
Professor and Chairman
Department of Cell, Viral, and Molecular Biology
University of Utah Medical Center
Salt Lake City, Utah

TABLE OF CONTENTS

Volume I

TABLE OF CONTENTS

Volume II

TABLE OF CONTENTS

Volume III

Chapter 1

NATURAL HISTORY OF THE RHABDOVIRUSES OF VERTEBRATES AND INVERTEBRATES

F. Brown and Joan Crick

TABLE OF CONTENTS

I. INTRODUCTION

Rhabdoviruses are found in many vertebrates, invertebrates, and plants. Because of this wide host range there has been much speculation for many years as to whether viruses found in one kind of host will also infect other kinds of host. Some members have, in fact, been shown to replicate in both vertebrates and invertebrates and others in both plants and invertebrates. For example, vesicular stomatitis virus will infect several vertebrate hosts and also multiply in *Aedes* mosquitoes.[1-3] Furthermore, Lastra and Esparza[4] have shown that vesicular stomatitis virus will grow in leafhoppers, which are the natural vector of maize mosaic virus, a plant rhabdovirus. It has been suggested that some rhabdoviruses which infect plants may also infect vertebrates,[5] but the evidence for this seems to be slight. In this chapter we have confined our attention to those members infecting vertebrates and invertebrates. The plant viruses are reviewed by Francki and Randles[6] in Chapter 7 of Volume III.

The characteristic bullet shape of the rhabdoviruses means that they are easily recognized in the electron microscope. So far about 30 such viruses have been found in vertebrates but only about one third have been characterized by methods other than electron microscopy. Two genera, *Vesiculo* and *Lyssa,* have been proposed on the basis of serological interrelationships.[7] The *Vesiculo* genus includes the different serotypes of vesicular stomatitis virus and also Piry, Chandipura, and Isfahan viruses (Table 1). Although other viruses are present in the list given by Fenner,[7] we consider that there is insufficient evidence for their inclusion in the *Vesiculo* genus at this stage.

By far the best characterized member of the group is vesicular stomatitis virus. Historically this has been because vesicular stomatitis has a close clinical relationship with the more economically important foot-and-mouth disease. In addition to the two classical serotypes of vesicular stomatitis virus, Indiana and New Jersey,[8] which have been known for more than 50 years, three other closely related viruses have been found during the last 15 years.[9-11] These are the Cocal, Argentina, and Brazil (Alagoas) strains, which are related to the Indiana serotype. Piry,[12] Chandipura,[13] and Isfahan[14] viruses have also been allotted to the *Vesiculo* group but the serological relationship of these three viruses to each other and to the other *Vesiculo* viruses is much less pronounced.

The *Lyssa* group, listed in Table 2, includes rabies virus which is by far the most important rhabdovirus infecting vertebrates. This virus will infect all warm-blooded animals and the disease is almost invariably fatal. In recent years three viruses related serologically to rabies virus have been isolatee in Africa. These are Mokola,[15] Lagos bat,[16] and Duvenhage viruses[17] and they can be regarded as new serotypes of rabies virus. They could be important epidemiologically because animals vaccinated against rabies are often poorly protected against the related viruses. Two other viruses, Obodhiang[18] and Kotonkan,[19] also found in Africa, are distantly related to Mokola virus.

Far less is known about the other rhabdoviruses. A list of these viruses and the hosts in which they have been found are given in Table 3. Apart from bovine ephemeral fever virus and the viruses which infect fish and which are clearly of great and increasing economic importance, the other members are of interest mainly because of their possible relatedness to the remainder of the family. Bovine ephemeral fever virus is the subject of Chapter 8 Volume III[20] and the fish viruses are described in Chapter 9 in Volume III[21] and will not be considered here.

II. VESICULAR STOMATITIS VIRUS

A. Introduction

Vesicular stomatitis virus is the prototype of the family Rhabdoviridae. The virus

TABLE 1

The Members of the *Vesiculo* Genus of the Rhabdoviridae and their Geographical Distribution

Virus	Species from which virus has been isolated	Species in which specific antibody has been found in serological surveys	Geographical distribution
Vesicular stomatitis			
Indiana	Cattle, horses, horseflies, man, mosquitoes, pigs, sandflies	Birds, cattle, dogs, horses, man, pigs	North and South America
Cocal	Mites, rodents	Horses, rodents	Trinidad, Brazil
Argentina	Horses	Horses	Argentina
Brazil	Cattle, horses	Cattle, horses, man	Brazil
New Jersey	Cattle, horses, man, pigs	Cattle, deer, dogs, horses, man, pigs	North and South America
Piry	Man, opossums	Bats, edentates, man, marsupials, monkeys, pigs, rodents	Brazil
Chandipura	Hedgehogs, man, sandflies	Buffaloes, cattle, camels, donkeys, goats, horses, man, monkeys, sheep	India, Nigeria
Isfahan	Sandflies	Gerbils, man	Iran

TABLE 2

The Members of the *Lyssa* Genus of the Rhabdoviridae and their Geographical Distribution

Virus	Species from which virus has been isolated	Species in which specific antibody has been found in serological surveys	Geographical distribution
Rabies	Many warm-blooded animals	[a]	Worldwide
Lagos bat	Bats	None	Nigeria, Central African Republic
Mokola	Man, shrews	Bats, birds, cattle, goats, man, pigs	Nigeria, Cameroons
Duvenhage	Man	[b]	South Africa
Kotonkan	Culicoides	Bats, cattle, horses, hedgehogs, rodents, sheep	Nigeria
Obodhiang	Mosquitoes	None[c]	Sudan

[a] Most species infected with rabies virus die. However some species can become inapparently infected and produce antibody.[67,67a]

[b] No information.

[c] None — survey made but no antibody detected.

can infect horses, pigs, and cattle, and is clinically indistinguishable from foot-and-mouth disease, vesicular exanthema, and swine vesicular disease, all of which are of economic importance to the farming industry. Consequently vesicular stomatitis virus has received a great deal of attention. It is also an excellent model for studying a variety of problems in molecular virology.

B. The Disease

The initial signs of vesicular stomatitis in farm animals are the inability to eat, the formation of froth on the lips, and the dribbling of saliva. In the mouth small blisters appear which erode quickly and develop into ulcers. Vesicular lesions also appear in the skin of the coronary band of the feet. In horses all four feet are often affected

TABLE 3

Unclassified Members of the *Rhabdoviridae* and their Geographical Distribution

Virus	Species from which virus has been isolated	Species in which specific antibody has been found in serological surveys	Geographical distribution	Ref.
Bovine ephemeral fever	Cattle, culicoides	Antelopes, cattle	Africa, Australia, Japan, Middle East	20
Marco	Lizards	—	Brazil	47, 143
Kern Canyon	Bats	—	California	144
Klamath	Rodents	—	California	144
Barur	Bats, mosquitoes, rodents, ticks	—	Kenya, India	47, 143
Flanders	Birds, mosquitoes	—	North America	145
Hart Park	Birds, mosquitoes	—	North America	144, 146
Mount Elgon bat	Bats	—	Kenya	147
Joinjakaka	Mosquitoes	None	New Guinea	47
Navarro	Birds	—	Columbia	47
Kwatta	Mosquitoes	None	Surinam	148
BeAn 157575	Birds		Brazil	143
Mossuril	Birds, mosquitoes	Baboons, man	Central and Southern Africa	149
Kamese	Mosquitoes	Man	Uganda	47
Sawgrass	Ticks	—	Florida	140
New Minto	Ticks	—	Alaska	143
Chaco	Lizards	None	Brazil	47
S-1643	Mosquitoes	—	Sarawak	143
BFN-3187	Mosquitoes	—	California	143
Oita 296	Bats	—	Japan	143
Kununurra	Mosquitoes	None	Australia	140
Egtved (viral hemorrhagic septicemia)	Trout	None	Europe	21
Sigma	Fruit fly	None	France	150
Infectious hematopoietic necrosis	Salmon, trout	None	United States	21
Spring viremia of carp	Carp	None	Europe	21
Pike fry disease	Pike	None	Europe	21
Rhabdovirus of eels	Eels	None	Japan	151
Rhabdovirus of grass carp	Grass carp	None		152
Rhabdovirus of entamoeba	*E. invadens, E. histolytica*	—	—	143
Rhabdovirus of blue crab	crab	—	Canada	153

whereas in cattle blisters on more than one site are rare.[5,8] This is the main clinical distinction between vesicular stomatitis and foot-and-mouth disease in cattle since generalization is much more common in the latter disease. The temperature of the animals, rises but falls rapidly to normal after rupture of the vesicles. The duration of sickness is about 2 weeks if there are no complications.[8] Inapparent infection also occurs judging by the presence of neutralizing antibody to the virus in the serum of farm animals. For example, all individual animals in a herd of 26 cattle in Wisconsin possessed virus-specific antibodies 1 month after natural infection although the disease had been observed in only half of the animals.[22]

In addition to its importance in farm animals, vesicular stomatitis virus will also infect man.[23] In man the disease usually resembles influenza, although there are reports of a cutaneous form of the disease from accidental infection of laboratory workers.[24] Inapparent infections also occur in man judging from the high antibody levels in the sera of people who had been in contact with the virus but had not experienced the disease.[25]

C. History

When four viruses are known which cause vesicular disease in farm animals it is clear that diagnoses made before adequate serological tests were available can only be regarded as circumstantial. However, of the four viruses, only vesicular stomatitis virus regularly causes disease in horses so that the early reports of vesicular stomatitis in this species can probably be regarded as authentic.

According to Hanson[26] the first reports of the disease in horses came from the U.S. in 1821 and there have been sporadic outbreaks in that country ever since. There is a good description of a disease in horses during the American Civil War[27] which suggests that vesicular stomatitis virus was the cause of the infection. "The artillery and cavalry required large numbers to cover losses sustained in battle, and on the march and by diseases. Both of these arms were deficient when they left Washington in September 1862. A most violent and destructive disease made its appearance at this time which put nearly 4000 animals out of service. Horses reported perfectly well one day would be dead lame the next, and it was difficult to foresee when it would end or what number would cover the loss. They were attacked in hoof and tongues. No one seemed able to account for the appearance of the disease. Animals kept at rest would recover in time but could not be worked."

Vesicular stomatitis is regarded as a disease of the Western Hemisphere since there are only three well-documented reports of its occurrence outside that area. In 1884 and 1897 stomatitis occurred in horses in South Africa[28] and in 1915 to 1917 a similar outbreak occurred among horses in France.[29] In the last outbreak cattle were also affected.[37] The European outbreak was traced to horses imported from Canada and the U.S. and it is likely that the outbreaks in South Africa had a similar origin. Later outbreaks of stomatitis in South Africa in 1934, 1938, and 1943, in which the disease was confined to cattle, may have been caused by one of the South African serotypes of foot-and-mouth disease virus which at that time would have been difficult to identify by serological methods. Other reports on the occurrence of the disease outside the Western Hemisphere are also of doubtful authenticity.

The first transmission experiments were done in 1925 in Texas and Indiana with material isolated from cattle with large lesions in their mouths and on their feet. The disease was transmitted to horses by inoculaton.[8] The infectious agent from the Indiana outbreak was isolated and preserved by animal transfer. It is known as the Indiana strain.

In the following year a serious outbreak of vesicular disease occurred in New Jersey, particularly in cattle. Cotton[8] showed that the causative agent was a virus immunologically distinct from the Indiana virus isolated in the previous year. These two antigenic types, Indiana and New Jersey, are the two classical serotypes and until 1963 they were the only serologically distinct strains of the virus that had been described. In that year, however, Jonkers et al.[9] recovered Cocal virus from mites collected from rice rats trapped in Trinidad following a rodent epizootic in that area and in Brazil. In neither instance was the virus associated with clinical stomatitis although strain specific antibodies were found in horses and in rodents. This virus is related to but different from the Indiana serotype and is quite distinct from the New Jersey serotype.[31] Two other viruses which are also related distantly to the Indiana serotype were recovered from outbreaks of disease in horses in Argentina[10] and in Alagoas, Brazil[11] and are the first isolates of the virus from the countries concerned. The affected animals in the Argentinian outbreak had severe vesicular lesions involving the tongue, gums, and lips and also the coronary band. In the outbreak in Brazil mules and horses were affected most frequently but cases occurred also in cattle and strain-specific antibodies were found in the serum of plantation workers who had complained of fever, headaches, and malaise at the time of the outbreak.

D. Present Distribution of the Disease

The disease is enzootic in Central America and the southern states of the U.S. but extends occasionally as an epizootic into the U.S., Canada, and South America. It reappears each year in the enzootic regions but occurs much less frequently in the epizootic regions. The New Jersey serotype has occurred more frequently than the Indiana serotype in recent years.[32]

E. Epizootiology

The epizootiology of vesicular stomatitis has been extensively reviewed by Hanson.[22,26] In these reviews he stressed that an understanding of the transmission of the disease requires a knowledge of the species that are susceptible under natural conditions. Clinical signs of vesicular stomatitis have been seen only in horses, cattle, pigs, and man. However, serological studies among wild animals in enzootic areas have shown that many species have contact with the virus in nature.[33-37] Some birds have also been shown to carry virus neutralizing antibody.[34,38] The possibility that bats could be carriers of the disease[22] was studied by Donaldson[37] who found that Cocal virus persisted in bats for 2 to 3 weeks after infection and could be transmitted to suckling mice by mosquitoes.

The geographical distribution of the epizootic disease and the limitation of the enzootic disease to tropical areas suggest that environmental factors are of considerable importance in its spread. Epizootic disease in the U.S. has been limited to the July to October period and the arrival of heavy frosts in October and November usually results in the termination of the outbreak. For an epizootic to occur it seems that the following conditions are important: (1) presence of a susceptible population, (2) temperatures above 20°C and a high humidity, and (3) presence of natural water. Treeless plains and deserts serve as barriers to the spread of the disease.

Three aspects of the epizootiology of the disease suggest that it is transmitted by a vector: (1) the seasonal incidence, (2) the ecological limitation, and (3) the rapidity and manner of spread. Thus the disease appears in summer and disappears shortly after frosts severe enough to kill insects. The disease also follows waterways but usually fails to spread along roads. For this reason epizootics usually occur in regions where cattle are pastured in woodland where streams or lakes are common rather than on open plains.

Hanson[22] considered that the horse fly was the most likely vector and the prevalence of this insect in Wisconsin is closely related to the extent of the 1937 and 1949 epizootics. Horse flies are also found in large numbers in other epizootic areas but are relatively rare in some of the regions which have so far escaped the disease. There is now evidence that vesicular stomatitis can be transmitted experimentally by horse flies and mosquitoes[40] and workers in Trinidad,[9,35] Panama,[41] and the U.S.[42] have isolated the virus from mites, biting gnats, and mosquitoes. Mussgay and Suarez[1] demonstrated that the virus can multiply in *Aedes aegypti* mosquitoes and this has been confirmed by other workers.[2,3,43] The virus can also grow in sandflies[44] and leafhoppers.[4] The demonstration that *Phlebotomus* flies not only become infected when fed the virus but can then transmit the virus transovarially to the next generation[45] provides evidence that these flies could serve as a reservoir between epizootics. The infection of insects by vesicular stomatitis virus may thus provide a means for the transmission of the disease between vertebrate hosts.

III. PIRY VIRUS

Piry virus was first isolated by Shope in 1960 from a spleen and liver sample of an opossum in Belém, Para, Brazil.[12] The virus has also been isolated from man and there have been a number of cases among laboratory staff who have worked with the virus.[46] The illness in man is one of sudden onset of fever but recovery is usually rapid.

Specific antibody has been found in man, monkeys, marsupials, pigs, edentates, water buffaloes, and rodents and also in a bat in Brazil.[47] No antibody was found in horses and cattle. The virus grows in *Aedes aegypti* mosquitoes[48] which suggests that these could act as vectors.

IV. CHANDIPURA VIRUS

Chandipura virus was first isolated from human sera obtained from patients with a febrile illness during an epidemic in India due to chikungunya and Dengue viruses.[13] The illness was similar to that found during infection with these two viruses. There was a sudden onset of fever without chills or rigor but with pain in several parts of the body. There was no rash or bleeding. The virus has also been isolated from sandflies in India and hedgehogs in Nigeria.[47]

Serological surveys have shown that the virus is widespread among humans in India and must have been present at least as early as 1955.[13] There is a wide variation in the proportion of sera that contain antibody between different areas of the country. Sera from horses, donkeys, cattle, sheep, goats, and monkeys have also been found to contain specific antibody to the virus.[47]

The evidence that the virus can be transmitted by mosquitoes under laboratory conditions may have implications regarding the spread of the disease in nature.[47]

V. ISFAHAN VIRUS

Isfahan virus was recovered from sandflies (*Phlebotomus papatasi*) during studies on the epidemiology of sandfly fever in Iran in 1975.[14] During a serological survey in three different provinces in Iran neutralizing antibody was found in the sera of humans and gerbils but none was found in the sera of sheep, goats, cattle, chickens, or pigeons. The prevalence of antibodies among humans residing in 15 different communities showed that the infection rate varied from 0 to 86% in the different areas. Of 33 gerbils tested, 26 showed the presence of antibody.[14]

VI. RABIES VIRUS

A. Introduction

It is less than 20 years since rabies virus was first visualized in the electron microscope[49-52] but the disease it causes has been known for at least 4000 years.[53] All warm-blooded animals, including man, are susceptible to the disease and once clinical signs develop death is virtually certain. No disease is more dreaded and it is still prevalent in many parts of the world.

B. The Disease

In many species the disease can exist in two forms, the more easily recognized *furious rabies* in which the brain is predominantly affected and the less obvious *dumb or paralytic rabies* in which the spinal cord is involved. The appearance of overt disease is usually preceded by a prodromal period in which there are a number of nonspecific symptoms of malaise. As the disease progresses, infection of cells of the limbic system occurs and with it changes in emotional behavior characteristic of rabies.[53-55] In man one of the most frequently observed symptoms, the inability to swallow liquids, has led to the disease being given the name of "hydrophobia" and descriptions of mad dogs and their association with the disease date from ancient times. Other small animals lose their natural timidity when rabid and attack other species indiscriminately.

Rabies virus is neurotropic and viremia plays little or no part in its spread throughout the body.[56] In experimental animals it has been shown that once the virus is deposited in the tissues it travels through the peripheral nerves to the spinal cord and thence to the brain. It multiplies in all parts of the brain and then moves outwards through the efferent nerves to virtually every organ of the body, including the salivary glands and hence to the saliva in which it is transmitted, usually by biting.[57-61]

Rabies is characterized by a long and variable incubation period, from 10 to 209 days in animals and up to 3 years in man (4 to 10% greater than 6 months).[56] The long incubation period has made postexposure therapy in man seem feasible and is also important in the spread and maintenance of the disease. Shope[62] considers that in dogs, for example, the long incubation period is responsible for the ease of introduction of rabies into areas free from the disease and also allows susceptible offspring to enter an infected population, thus ensuring continuation of the epizootic.

It is not known where or in what form the virus exists during the incubation period. Experiments with a number of strains of rabies and rabies-related viruses (see below) suggest that the virus may replicate in or near its site of entry into the body before entering the nervous system and that this phase may occupy a considerable time.[63] The role of the immune response in rabies pathology remains unclear; antibody levels in both the cerebrospinal fluid and serum are often very high after infection.[56]

Despite the widely held belief that rabies is invariably fatal, two well-documented cases of humans recovering from the disease have recently appeared, although it must be emphasized that virus was not recovered from either of these patients.[64,65] Among other animals, including dogs, there are many reports of long term inapparent infection accompanied by virus shedding.[66,67,67a] Bats in particular have been associated with this condition, a factor which may be important in their role as virus disseminators.[68-70] How such inapparent infections are maintained is not understood, although it is tempting to speculate that defective interfering (T) particles and/or interferon may be involved.

C. Transmission

Transmission usually occurs during the furious stage of the disease when the animal makes unprovoked attacks and so infects new hosts by biting. For obvious reasons, carnivores and small rodents are the animals usually involved. Infection may also occur by the licking of grazed or abrased skin or by contact of the mucous membranes with infected saliva. Aerosol infection is also possible; infection of man and animals in bat-infested caves has been recorded[71,72] and two recent cases of infection of laboratory workers have also been reported.[73,74] Fortunately, not every case of exposure to virus leads to the development of disease; the chances of human infection, for example, have been placed as low as 0.1% from minor wounds from infected animals but up to 60% following severe bites. Many other factors are also involved including the proximity of the bite to the brain and the species of attacking animal.[75]

D. Diagnosis

Rabid animals endanger man and his domestic stock so reliable diagnostic tests are crucial. Until 1903 no such test was available for rabies diagnosis but in that year Negri[76] described inclusions in the neurones of brains from rabid animals which he regarded as typical of rabies virus infection. These "Negri bodies" which consist of masses of ribonucleoprotein, are often very good indicators of the presence of rabies virus but are not seen in all cases and similar staining inclusions may be produced in other conditions.[77] Nevertheless, the test has been, and still is, regarded as a useful guide as to whether or not rabies is present, though it has now been largely superseded by the mouse inoculation test and the very rapid and more reliable fluorescent antibody test.[78] It should be remembered, therefore, that until these tests were available all cases of rabies were identified solely on the basis of clinical evidence.

E. Epizootiology

Rabies is maintained in two interrelated phases described as urban and sylvatic. In the former, domestic and stray dogs and cats are involved whereas in the latter infected wild animals transmit the disease. Even dog rabies was considered by Johnson[79] to be primarily a disease of wild canine species, domestic dogs having no role in the maintenance or spread of the disease before the 18th century, when industrialization began in Europe and America. The increase of urban rabies became more of a problem as cities increased in size; nowadays it is rigorously controlled in those parts of the world where the efficient application of public health measures is possible.[80]

Sylvatic rabies, on the other hand, presents a much more intractable problem and a constant threat to man and his livestock. Rabies epizootics in wild life have been reported since the Middle Ages but have only been extensively studied, chiefly in Europe and North America, since the current upsurge began. These epizootics are characterized in particular locations by the predominance of one or two carnivore species as the main victims and vectors[66,81] and the pattern of the epizootic in an area remains stable over many years. The red fox is probably the animal most susceptible to rabies[66] and it has been implicated in several major epizootics, including the present one in Europe. Since the Second World War, there have been large-scale agricultural and industrial change in Northern Europe, to which the fox has proved more adaptable than its natural competitors. Furthermore, the number of foxes increased during the war, presumably due to the suspension of hunting, so that the population became better suited for the maintenance and spread of rabies.[82] A few other animal species, including badgers, martens and deer, are regularly found to be infected, but these are regarded as having no significance in maintaining the current outbreak. The isolation of rabies-like viruses in rodents in Central Europe[83] is also considered to be of no epizoological importance.[84]

Several species of foxes have been recognized for many years as a major rabies reservoir[85] in the U.S. and Canada but unlike Europe, in these countries other wild animals are also important in maintaining and transmitting the disease (see Figure 1). For example the skunk has the major role in spreading rabies in the Midwest and in California[53,66,86] and is now the most important source of human infection in the U.S.[86] It is a most efficient disseminator of the disease since it has long periods of virus emission and high virus titers in the saliva. In Georgia and Florida the raccoon has been recently

RABIES Epizootic aspects

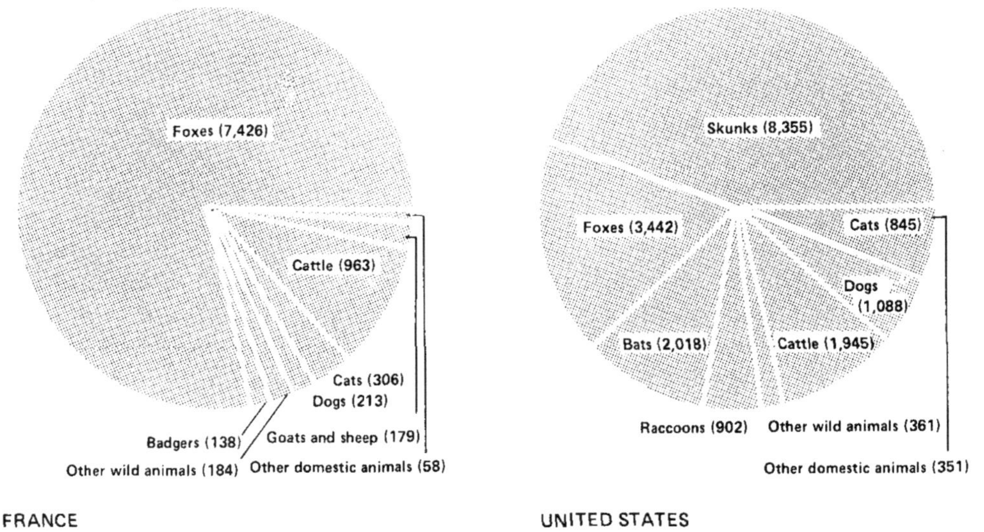

FIGURE 1. Number of confirmed cases of rabies in animals in France from March 1968 to December 1975 and in the U.S. from 1969 to 1973.[163] (From Taylor, D., *Vet. Rec.*, 99, 157, 1976. With permission.)

recognized as an important transmitter of the disease, though it is of little importance elsewhere in the U.S.[87]

Bats are of particular significance in the spread of rabies in the Americas. The association of vampire bats with the disease was definitely established in 1921 and 10 years later in Trinidad frugivorous bats were also shown to be able to transmit rabies.[68,70] Insectivorous bats (the only type found in the U.S.) were subsequently found to have a similar ability[53,68] and it has been suggested that the virus may persist in the brown fat of these animals during hibernation.[88,89] Bats are the most widespread of animal hosts in the U.S. and rabid bats have been isolated from most states.[90] In 1975 they accounted for 19% of all the virus isolates in the country.[53]

The importance that has been attached to the understanding of the rabies cycle in bats may be judged by the extensive treatment of the subject in *The Natural History of Rabies* edited by G.M. Baer, which appeared in 1975,[91] and it seems that both abortive infections and the asymptomatic carrier state may occur in certain species.[53] In addition to their ability to transmit the disease by biting, bats have been implicated in airborne infections as was dramatically illustrated by the deaths of two men from rabies after exploring the bat-infected Frio cave in Texas.[71] Sentinel animals subsequently placed in the cave and protected from direct contact with arthropods and other animals also contracted the disease.[71]

In Europe and Asia bat rabies appears uncommon and extensive searches in various parts of Africa have failed to reveal any rabid bats. It is therefore of particular interest that two of the rabies-related viruses, Lagos bat virus,[16] and Duvenhage virus,[17] both found in Africa, were associated with infected bats (see below).

Among other animals known to be important in maintaining rabies are the jackal and mongoose in parts of South Africa and the wolf in Southeastern Europe, the Middle and Near East, Mexico, and Northern Canada.[92] Rats are still regarded as a potential source of human infection in Lebanon and parts of North Africa,[93] and enzootic rabies in rodents has also been reported from Thailand.[52] However, there is no critical evidence that rodents will transmit the disease to man.[95] Birds are also susceptible to rabies but there is little evidence that they have any role in the epizootiology of the disease.[96]

F. History

Rabies has always been associated primarily with dogs and descriptions of the disease in these animals in Ancient Mesopotamia date from the Third Century B.C.,[79,97] and in China from ca. 555 B.C.[53,98] It was also known to the Greeks and Romans who recognized that the disease could be transmitted in the saliva of a rabid dog by biting. The Romans described the infectious material as a virus (the Latin for poison)[80] and Celcus in the first century A.D. appears to have been the first to associate rabies in animals with hydrophobia in man.[80,96]

From this time until the 19th century the pattern of the disease in Europe appears to have been of outbreaks in wild and domestic dogs followed by infection in other animals and in humans. It is probable that these outbreaks may have been preceded by epizootics in foxes.[62,99] There are many accounts of people and domestic animals being attacked by rabid foxes and other wild animals;[80,96,99,100] for example, Johnson[79] and Irvin[96] cite wolf rabies in Western Europe in 1271. At least seven epizootics are known to have occurred in foxes between 1803 and 1928.[99,101] One major outbreak in which this species was implicated occurred during the period of the Napoleonic Wars, but after 1880 the disease in wildlife appeared to wane. However, rabies continued to occur in domestic animals, particularly dogs, until the turn of the century.

In view of the fear engendered by the disease, it is perhaps not surprising that it became common to adopt many curious and superstitious practices to avoid its inevitable outcome.[53,80,98] Yet some scientific progress towards its understanding was also made. Several workers, of whom the first was Zinke,[100] succeeded in transmitting the disease to healthy dogs by inoculation of infected saliva, while Galtier[102] by intravenous inoculation of the same material was able to protect sheep against subsequent challenge. Galtier also showed that the rabbit was a useful experimental animal for the study of rabies.[103] He thus paved the way for Pasteur and his co-workers to develop virus attenuated for dogs but having a uniform (fixed) incubation period in rabbits and to prepare the first nervous tissue vaccines for immunization of dogs and man.[104]

The experiments of Zinke and Galtier[100,101] convinced authorities that rabies was an infectious disease and that the control of dogs might lead to its elimination. Early in the 19th century quarantine regulations, elimination of stray dogs, and muzzling were introduced in Scandinavia and by 1826[101] this area was free from dog rabies. Dog control regulations were also applied successfully to other urban centers in Europe, but these areas frequently became reinfected after a few years.[101]

In Britain, geographically isolated from mainland Europe by the sea and apparently free from wild life rabies, the authorities were much slower to introduce measures for dog control. Although the disease had been present in the country for most of recorded history and in the 19th century was presenting a serious problem in industrial midland towns,[79,80,92,99] no attempt was made to deal with it on a national basis until 1831, when a bill was introduced in Parliament to "prevent the spreading of canine madness." The bill did not become law but local authorities were later given various powers to control dogs (e.g., muzzling, destruction of strays, etc.).[80,92,99] It was not until 1897 that legislation requiring a quarantine period for infected animals was introduced, yet by 1903 Britain was free from rabies. Apart from an outbreak between 1918 and 1922, probably caused by the illegal importation of dogs by servicemen returning from the First World War, the country remained free from the disease until 1969 and 1970. In each of these years a dog became rabid after being released from the statutory 6 months in quarantine. Fortunately, neither of these incidents led to cases of secondary infection and despite the importation of about 5000 dogs and cats per year into the country, Britain has returned to its rabies-free status.[92,95,99]

With the exception of Scandinavia and Britain, complete elimination of rabies from

Western Europe proved virtually impossible because of its persistence in the wildlife. Despite its apparent disappearance from some areas for long periods, it has emerged at intervals with the consequent spillover into dogs and other domestic animals. Two major resurgences of the disease in Central and Western Europe have been associated with the First and Second World Wars. As Schoop[105] has pointed out however, in the first, from 1918 to 1922, dogs were mainly implicated, while in the second, which started in 1939 and is still continuing (see below) foxes are the main cause of dissemination.

At the beginning of the Second World War, numerous cases of rabies in foxes and badgers were reported in Poland and in 1948 the disease reached Czechoslovakia and Austria and by 1950 was present in West Germany. It was temporarily halted at the Rhine in 1960, but has since continued to spread westward. Its presence in Belgium and Luxembourg was confirmed in 1967 and in 1968 it entered France. In Germany and in Denmark which became reinfected in 1964, the disease has been controlled but not eradicated by drastic reduction of the fox population but in France this procedure has been less successful and the disease has continued to spread westward.[99]

It is somewhat surprising that in Europe, where the first rabies vaccines were developed, vaccination was not used until comparatively recently in the control of canine rabies. In the period following the upsurge of the disease after the First World War, Hungary did in fact vaccinate dogs in increasing numbers each year, so that by 1939 the disease had been eliminated from that country.[106] Even in the present epizootic and despite the availability of more reliable vaccines, several authorities in Western Europe were reluctant to use veterinary vaccination as a control weapon though it is now becoming more generally accepted.[99]

Rabies was unknown in the American Continent until the mid-18th century, when it was probably imported by European settlers taking dogs with them. The first reports of rabies in the Western Hemisphere were of an outbreak in dogs and cattle in Barbados in 1741[80] and in dogs in Virginia in 1753.[62] The first major epizootic in North America was reported between 1768 and 1771 when foxes and dogs spread the disease to pigs and other domestic animals, and by 1785 dog rabies was widespread in all the colonial states. Wildlife rabies was also recognized in the eastern states in the 18th Century and after the Civil War dog rabies was prevalent in most of the U.S.[92] Canine control measures were introduced unsuccessfully at the beginning of the 20th century and it was not until the 1950s when a policy of widespread dog vaccination was adopted that urban rabies was successfully controlled.[80,107]

In South America a major epizootic was reported in Peru between 1803 and 1808 when man, domestic animals, and even fowl were said to have been infected.[80] Although the epizootic subsided, the disease remained endemic. Rabies was also reported in Argentina in 1806, Uruguay in 1807, and Chile in 1838.[108] The disease also occurred in Colombia in the early 19th century where intensive measures to control dogs were taken in order to curtail the outbreak.[92]

The problem of rabies transmitted by bats, which is almost exclusively confined to the Americas (see above), constitutes a recent chapter in the history of the disease.[66,92] There are a number of reports by the early Spanish colonists of vampire bats attacking other animals and men, and of deaths caused by such bites, but it is only since 1921 that vampire bats have definitely been associated with rabies, particularly the paralytic or dumb form seen in cattle.[70,109] Derriengue is the common Spanish name for this form of the disease in cattle, the history of which was described by Johnson.[110] The first diagnosis of rabies in a non-hematophagous bat was made in Trinidad in 1931 and since then regular diagnoses of rabies in insectivorous and frugivorous bats have been made in various parts of the American continent.[68]

A brief historical review of rabies in Asia was given by Tierkel in 1971.[97] In the mid-19th century, several reports of the disease in dogs in China were made by resident Europeans and the disease in man was well known to Chinese physicians.[80] However, the study of the disease in the East did not really begin until the early 1900s when Pasteur Institutes were established in India, Iran, Vietnam, and Indonesia.[97] In India, shortly after this time Semple[111] produced the first fully inactivated rabies vaccines and in 1920 Japan became the first country to practice the large-scale vaccination of dogs.[112] This, combined with other control measures, led to the final eradication of the disease in that country in 1956, after a massive outbreak of canine rabies from 1948 to 1950 in the Kanto district which includes the Tokyo metropolis.[113]

Rabies has been known in North Africa since ancient times and was mentioned by the School of Alexandria.[98] In 1858 preventive measures were introduced in Algeria to control the extensive spread of the disease, which a few years later was reported as enzootic in Ethiopia.[80] Curiously, travelers such as Burrow in 1806 and Livingstone in 1857 commented on the complete absence of the disease[92] and in 1893 the first outbreak known to have occurred in South Africa was caused by a dog imported from England.[92] This paucity of reports of rabies in the African continent is particularly interesting in view of the fact that rabies may have originated there.[62]

G. Present Distribution of the Disease

Rabies is still present in most parts of the world except Australia, New Zealand, and Antarctica, and countries such as Britain and Japan which have succeeded in eliminating the disease remain free from it because of their geographical isolation and strict control measures.[92,95,99,113-117] In other countries such as Denmark[118] which have freed themselves of rabies, the disease has reappeared after variable periods of time. In 1959 the Veterinary Public Health Unit of the World Health Organization (WHO) began regular surveys of rabies[118] which since 1969 have been supplemented by a monthly surveillance of the disease in the Americas carried out by the Pan American Zoonoses Center in Buenos Aires.[119] In addition, the yearly incidence of rabies is published in the annual Animal Health Year Book, published jointly by WHO, Food and Agriculture Organization (FAO), and the Office International des Epizooties (OIE).[32] Unfortunately the number of countries sending statistics to WHO is declining;[120] furthermore, the accuracy of reporting varies from country to country, so that the world picture presents "an impression rather than a precise record."[93] The epidemiology of rabies has been reviewed recently by Kaplan,[46] Bisseru,[92] Turner,[93] Irwin,[96] Kaplan,[114] Abdussalam and Botton,[121] and Johnson.[122]

Compared with many other diseases which occur on a global scale, human deaths and animal losses from rabies may not be striking although they are far from insignificant. Most of the cost is involved in preventive measures, from the rounding up of stray animals to the provision of massive veterinary vaccination programs and postexposure treatment for man.[114] However, it is estimated that in the area between tropical Mexico and North Argentina the scarcely credible number of 1 to 2 million animals die each year from vampire bat rabies, at an annual cost to the farming industry of $80 to 250 million,[123] and in the last few years appreciable losses of cattle infected with rabies have also been reported from Canada, U.S., and South Africa, and in Europe from France, Germany, and the U.S.S.R.[66,118]

Animal species currently important in the maintenance and spread of sylvatic rabies are discussed in The Natural History of Rabies, Vol. II, and the spread of fox rabies in Europe has been discussed in at least three lengthy articles since 1976.[124-126] However, foxes are not usually considered to be a direct hazard to man[127] although 10 of the 38 people who died from rabies in Europe between 1966 and 1973 were reported to have been bitten by this animal.[93]

Since 1946, 13 human deaths from rabies have occurred in Britain, 5 of these since 1975.[128] All of these contracted the disease outside Britain and ten are known to have been infected by dogs. The dog still remains the animal most likely to infect man.[93,107,118] Despite the decline in the last 20 years or so in the incidence of infection in dogs in the U.S., it is mainly this species whose bites cause 30,000 postexposure treatments of man to be given each year.[129] Kappus has pointed out that this species should still be regarded as important in rabies transmission to humans in North America, chiefly in rural areas where there is contact with infected animals.[130]

In Africa, particularly in the north and west of the continent, Asia, Southeast Europe, and Central and South America, urban rabies remains of overriding importance and it is in these areas that most human exposures occur.[118] In India and the Philippines, the situation is of particular concern.[93,107,118] It has been estimated that in India there are more than 18,000 deaths from rabies each year, while more than 3 million people receive postexposure vaccination[131] with nervous tissue vaccines,[111] a procedure which itself entails considerable discomfort and no little risk to the patient.[132]

VII. RABIES-RELATED VIRUSES

A. Introduction

Although minor antigenic differences between isolates have been reported, rabies virus was regarded until recently as a single serological entity[133] embracing not only the classical virus strains but those associated with Derriengue[110] (vampire bat rabies), Oulo Fato,[66,92,96,134] a nonfatal encephalitis of dogs in West Africa, and Arctic rabies.[96,135] The viruses isolated recently from rodents in Czechoslovakia and West Germany are also considered to be strains of rabies virus.[83]

However, in the last 20 years or so several new viruses have been isolated which have been shown to have more distant serological and antigenic relatedness to rabies virus and to each other. Together with rabies virus itself, they are considered at present to be members of a separate genus within the Rhabdoviridae. The subgroup to which they have been assigned (Table 2) is designated the Lyssa group[7] (lyssa = madness).

Two of these viruses, Mokola[15] and Duvenhage,[17] have been associated with clinical disease[17,136-138] in man and each has caused a fatal infection.[17,138] There are some doubts as to whether Nigerian horse virus, associated with a clinical condition known as "staggers",[139] should be considered as a strain of rabies virus or a separate member of the Lyssa group. For this reason, this virus has not been included in Table 2.

B. History and Distribution

The rabies-related viruses have all been isolated in Africa, a finding which supports Shope's idea that rabies virus itself originated in that continent.[62] Lagos bat virus[16] was obtained in 1956 from the brain of a fruit bat (*Eidolon helvum*) on Lagos Island in Nigeria and in 1974 was also found in *Micropteropus pussilus* in the Central African Republic.[140] Three isolations of Mokola virus were made in 1968 from the viscera of shrews (*Crocidura* sp.) captured near Ibadan in Nigeria[15] and a new isolate from the same species was made in 1969.[136] A further isolate from shrews was made in 1974 in the Cameroons.[140] Mokola virus was also obtained in 1968 and 1971 from two young children suffering from central nervous system disease.[136-138] The second child died.[136,138] However, there is no evidence of a widespread distribution of Mokola virus antibody in sera from humans or domestic animals, birds, bats, or shrews in Nigeria.[136,141]

The virus nevertheless can infect adults, as demonstrated by the recent infection of a laboratory worker in West Germany. Fortunately, the individual concerned had a high level of rabies virus neutralizing antibody prior to infection, which may have aided his recovery.[142]

Duvenhage virus, also showing serological relatedness to rabies virus, was isolated in 1970 from the brain of a man in the Republic of South Africa who developed a rabies-like illness after being bitten on the lip by a bat.[17]

The two remaining viruses tentatively assigned to the Lyssa group are of particular interest since they are insect-borne. The first, Obodhiang virus, was isolated on three separate occasions from unengorged *Mansonia uniformis* mosquitoes in the Sudan,[18] while kotonkan virus was obtained from a pooled sample of *Culicoides* midges in Ibadan in Nigeria.[19] While Obodhiang virus has not been associated with any clinical disease, kotonkan virus may be the cause of a disease in cattle in Nigeria which resembles bovine ephemeral fever, and neutralizing activity for kotonkan virus has been found in that country in a high percentage of cattle sera and also in sera from rats, sheep, horses, and hedgehogs.[19]

VIII. OTHER VIRUSES

While the natural histories of vesicular stomatitis virus and rabies virus have been studied extensively, there is much less information on the other members of the rhabdovirus group found in vertebrates and invertebrates. This situation reflects the importance of the first two viruses. Indeed, with the exception of bovine ephemeral fever virus and the fish rhabdoviruses, most of the viruses listed in Table 3 have not been found to be associated with disease in vertebrate hosts under natural conditions although they produce encephalitis in the highly artificial situation of intracerebral inoculation of suckling mice. Nevertheless, specific antibody to several of the viruses has been found in species living under natural conditions where there has been no description of the occurrence of disease. It seems likely that disease among vertebrate hosts may be easily overlooked unless it affects man and his domestic animals. It is also possible that, even in the absence of overt disease, sub-clinical infection may occur. Apart from Flanders and Hart Park, Kwatta and BeAn 157575, and Mossuril and Kamese viruses, none of the viruses listed in Table 3 are serologically related to each other nor to any of those in Tables 1 and 2.

Although it was considered some years ago that the Marburg agent was a possible member of the rhabdovirus group, neither Marburg nor the related Ebola virus has been considered in this review because there is insufficient evidence to support their inclusion in the group.

IX. DISCUSSION

The rhabdoviruses are found in a wide variety of hosts and cause several important diseases of man, domestic animals, fish, and plants. Their similar physico-chemical characteristics point to a family which is narrowly defined, in contrast to their diverse ecological and biological properties. Despite their economic importance, however, the natural history of these viruses has not been fully investigated; yet it is on such studies that appropriate control measures must be based.

For example, most vesicular stomatitis outbreaks appear to originate in semitropical America, yet at present we have insufficient knowledge as to how the virus is maintained between outbreaks and whether there is a wild life reservoir of the disease. In an extensive serological survey in Panama, Tesh and his co-workers[36] found evidence of widespread infection with both serotypes of the virus among humans, cattle, and many species of wild animals. Since it is known that cattle can be infected without showing signs of infection it is possible that this species may act as a reservoir for the virus. Viremia in vesicular stomatitis in cattle, horses, and rodents is short-lived, so there will be limited opportunity for the infection of hematophagous insects.[154] However, the seasonal incidence and the rapidity and manner of spread of the disease suggest that the virus is transmitted by a vector. The isolation of the virus from various species of insects and the demonstration that virus replication may occur in some of them provide evidence that insect vectors may be involved in transmitting the disease and/or maintaining it between outbreaks.

The idea of Johnson and his colleagues[5] that vesicular stomatitis may be a plant virus which also replicates in insects and spills over into the vertebrate population is highly imaginative, but requires firm experimental support before it can be seriously considered.

Of the other vesicular viruses, both Piry and Chandipura virus are known to infect man and specific antibody has been found in several different vertebrate species, including domestic animals.[47] As far as the authors are aware there has been no attempt to determine whether either of the viruses will produce overt disease in domestic animals or only inapparent infection. These seem to be points worthy of attention. Isfahan virus on the other hand, which appears to be more closely related to the New Jersey serotype of vesicular stomatitis virus than are Piry and Chandipura viruses[155] apparently infects only humans and gerbils. Unfortunately, the serological surveys of domestic animals carried out in Iran did not include horses or mules. It would be interesting to know whether these species can be infected by Isfahan virus, and whether a clinical condition similar to vesicular stomatitis could be induced by experimental infection.

The problems relating to the spread and maintenance of rabies are quite clearly different from those pertaining to the *Vesiculo* viruses. Urban rabies can be controlled but in countries where rabies is established in the wild life it is virtually impossible to eradicate it and the only effective way of protecting man and his domestic animals, apart from vaccination, is the selective culling, where practicable, of the appropriate species. Once again, a clear understanding of the natural history of the disease is required.

The current worldwide upsurge of sylvatic rabies is of such importance that it is being studied on an international basis.[118,119] It has become clear that populations of infected animals such as foxes, skunks, raccoons, and bats may exist in certain areas with very little transmission between the different species. The reasons for this are not understood nor are all the behavioral changes which occur in different infected animals well-documented. Most infected animals secrete virus in their saliva, but the stage of the disease when this occurs is not known for all species. Abortive, latent, and persistent infections are not uncommon in certain species and in animals such as vampire bats which live in colonies these infections appear to be important in the maintenance of the disease. Information on all these aspects of rabies is accumulating as more resources and personnel are provided.

Ten years ago Shope and his colleagues showed that rabies virus could no longer be regarded as a single serological entity when they demonstrated that Lagos bat virus and Mokola virus, isolated in Africa, were related to rabies virus.[15] The interrelationships were demonstrated in both serum neutralization and complement fixation tests and Schneider and Schoop[156] suggested that the two new viruses, together with Nigerian horse virus,[139] should be regarded as new serotypes of rabies. Schneider and Schoop considered that the rabies-related viruses shared a common antigen on the ribonucleoprotein which was responsible for the cross-reactivity observed in complement fixation and immunofluorescence tests.

The results of cross-neutralization tests within the Lyssa group, however, seem to vary between laboratories and cross-protection tests to determine antigenic relationships between rabies, Lagos bat, and Mokola viruses have also proved difficult to interpret.[157,158] There seem to be difficulties such as interference and prozone effects inherent in the tests.[158]

Obodhiang and kotonkan viruses isolated from insects have been shown by Bauer and Murphy[159] and by Buckley[160] to be more closely related to Mokola virus than to the rest of the *Lyssa* viruses. Both groups of workers have pointed to the fact that these viruses, unlike the other *Lyssa* viruses, will replicate in *Aedes aegypti* mosquitoes, and in vitro in *Aedes albopictus* cells. Furthermore Buckley was able to show quite clearly in plaque reduction tests that Mokola virus shared some surface antigens not only with rabies and Duvenhage viruses[17] (i.e., there was some cross neutralization) but also with Obodhiang and kotonkan viruses. She regarded Mokola virus therefore as a biological bridging agent and suggested that in the rabies virus serogroup there may be an interesting evolutionary model in which vertebrate viruses may have evolved through viruses with arthropod-vertebrate cycles from arthropod-borne virus ancestors. Bauer and Murphy[159] on the other hand drew attention to the cone-shape rather than the bullet-shape of Obodhiang and kotonkan viruses which they likened to that of bovine ephemeral fever virus.[20] The latter virus also occurs in Nigeria. The fact that kotonkan virus may be responsible for a disease in cattle in Northern Nigeria similar to that caused by bovine ephemeral fever virus was also considered significant.[19]

Examination of Klamath,[161] spring viraemia of carp, and pike fry disease viruses[162] shows that they have a morphology similar to rabies virus. The viruses causing viral hemorrhagic septicemia and infectious hemorrhagic necrosis of fish are also similar to rabies virus in their protein composition.[162] Such observations emphasize the need for more fundamental work in examining all possible interrelationships of viruses within the Rhabdoviridae, regardless of their origin.

ACKNOWLEDGMENTS

We wish to thank Mrs. P. Thomas and Mrs. P. Scott for their considerable help in trying to keep the authors sane during the preparation of this paper. We are also grateful to the Editor of the Veterinary Record for permission to reproduce Figure 1.

REFERENCES

1. **Mussgay, M. and Suarez, O.,** Multiplication of vesicular stomatitis virus in *Aedes aegypti* (L.) mosquitoes, *Virology*, 17, 202, 1962.
2. **Bergold, G. H., Suarez, O. M., and Munz, K.,** Multiplication in and transmission by *Aedes aegypti* of vesicular stomatitis virus, *J. Invertebr. Pathol.*, 11, 406, 1968.
3. **Liu, I. K. M. and Zee, Y. C.,** The pathogenesis of vesicular stomatitis virus, serotype Indiana, in *Aedes aegypti* mosquitoes. I. Intra-thoracic injection, *Am. J. Trop. Med. Hyg.*, 25, 177, 1976.
4. **Lastra, J. R. and Esparza, J.,** Multiplication of vesicular stomatitis virus in the leafhopper, *Peregrinus maidis* (Ashm.), a vector of a plant rhabdovirus, *J. Gen. Virol.*, 32, 139, 1976.
5. **Johnson, K. M., Tesh, R. B., and Peralta, P. H.,** Epidemiology of vesicular stomatitis virus: some new data and a hypothesis for transmission of the Indiana serotype, *J. Am. Vet. Med. Assoc.*, 155, 2133, 1969.
6. **Francki, R. I. B. and Randles, J. W.,** Rhabdoviruses infecting plants, *Rhabdoviruses*, Vol. III, Bishop, H. D. L., Ed., CRC Press, West Palm Beach, to be published, chap. 7.
7. **Fenner, F.,** Classification and nomenclature of viruses, *Intervirology*, 7, 42, 1976.

8. **Cotton, W. E.**, Vesicular stomatitis, *Vet. Med. (Kansas City, Mo.)*, 22, 169, 1927.
9. **Jonkers, A. H., Shope, R. E., Aitken, T. H. G., and Spence, L.**, Cocal virus, a new agent in Trinidad related to vesicular stomatitis virus, type Indiana, *Am. J. Vet. Res.*, 25, 236, 1964.
10. **Garcia Pirazzi, A. J., Coggiano, C. H., and Alonso Fernandez, A.**, *Publicación Tecnica No. 2 CA-NEFA*, Ministry of Agriculture, Buenos Aires, Argentina, 1963.
11. **Andrade, C. M.**, Studies on vesicular stomatitis in Brazil. Epidemiological inquiry in the Brazilian population, in *Global Impacts of Applied Microbiol.*, 4th Int. Conf., São Paulo, Brazil, 1973, p. 32.
12. **Taylor, R. M.**, Complete Catalogue of Arthropod-Borne Viruses of the World. A Collection of Data on Registered Arthropod-Borne Animal Viruses, U.S. Public Health Service, Publ. No. 1760, U.S. Government Printing Office, Washington, D.C., 1967.
13. **Bhatt, P. N. and Rodrigues, F. M.**, Chandipura: a new arbovirus isolated in India from patients with febrile illness, *Indian J. Med. Res.*, 55, 1295, 1967.
14. **Tesh, R., Saidi, S., Javadian, E., Loh, P., and Nadim, A.**, Isfahan virus, a new Vesiculovirus infecting humans, gerbils, and sandflies in Iran, *Am. J. Trop. Med. Hyg.*, 26, 299, 1977.
15. **Shope, R. E., Murphy, F. A., Harrison, A. K., Causey, O. R., Kemp, G. E., Simpson, D. I. H., and Moore, D. L.**, Two African viruses morphologically and serologically related to rabies virus, *J. Virol.*, 6, 690, 1970.
16. **Boulger, L. R. and Porterfield, J. S.**, Isolation of a virus from Nigerian fruit bats, *Trans. R. Soc. Trop. Med. Hyg.*, 52, 421, 1958.
17. **Meredith, C. D., Rossouw, A. P., and van Praag Koch, H.**, An unusual case of human rabies thought to be of chiropteran origin, *S. Afr. Med. J.*, 45, 767, 1971.
18. **Schmidt, J. R., Williams, M. C., Lule, M., Mivule, A., and Mujomba, E.**, Viruses Isolated from Mosquitoes Collected in Southern Sudan and Western Ethiopia, Rep. No. 15, East African Virus Research Institute, Entebbe, Uganda, 1965.
19. **Kemp, G. E., Lee, V. H., Moore, D. L., Shope, R. E., Causey, O. R., and Murphy, F. A.**, Kotonkan, a new rhabdovirus related to Mokola virus of the rabies serogroup, *Am. J. Epidemiol.*, 98, 43, 1973.
20. **Della-Porta, A. J. and Snowdon, W. A.**, Bovine ephemeral fever virus, *Rhabdoviruses*, Vol. III, Bishop, D. H. L., Ed., CRC Press, West Palm Beach, to be published, chap. 8.
21. **Roy, P.**, Fish rhabdoviruses, *Rhabdoviruses*, Vol. III, Bishop, D. H. L., Ed., CRC Press, West Palm Beach, to be published, chap. 9.
22. **Hanson, R. P.**, The natural history of vesicular stomatitis, *Bacteriol. Rev.*, 16, 179, 1952.
23. **Hanson, R. P., Rasmussen, A. F., Brandly, C. A., and Brown, J. W.**, Human infection with the virus of vesicular stomatitis, *J. Lab. Clin. Med.*, 36, 754, 1950.
24. **Schoening, H. W.**, U.S. Department of Agriculture, Washington, D.C., 1951; quoted in **Hanson, R. P.**, The natural history of vesicular stomatitis, *Bacteriol. Rev.*, 16, 179, 1952.
25. **Brody, J. A., Fischer, G. F., and Peralta, P. H.**, Vesicular stomatitis in Panama. Human serologic patterns in a cattle raising area, *Am. J. Epidemiol.*, 86, 158, 1967.
26. **Hanson, R. P.**, Vesicular stomatitis, in *Diseases of Swine*, Dunne, H. W. and Leman, A. D., Eds., Iowa State University Press, Ames, 1974, p. 308.
27. **McClellan, G. B.**, Report on the Organization of the Army of the Potomac and of its Campaigns in Virginia and Maryland, July 26, 1861 to November 7, 1862, The Times Steam Printing House, Chicago.
28. **Theiler, S.**, Eine contagiöse Stomatitis de Pferde in Süd-Afrika, *Dtsch. Tieraerztl. Wochenschr.*, 9, 131, 1901.
29. **Jacoulet, M.**, Au sujet d'une stomatite erosive de nature indeterminee (chez le cheval), *Bull. Soc. Cont. Med. Vet.*, 68, 576, 1915.
30. **Moussu, G.**, Discussion, *Bull. Soc. Cont. Med. Vet.*, 68, 578, 1915.
31. **Federer, K. E., Burrows, R., and Brooksby, J. B.**, Vesicular stomatitis virus — the relationship between some strains of the Indiana serotype, *Res. Vet. Sci.*, 8, 103, 1967.
32. Animal Health Year Books 1960—1975, FAO, WHO, OIE.
33. **Karstad, L. H., Adams, E. V., Hanson, R. P., and Ferris, D. H.**, Evidence for the role of wildlife in epizootics of vesicular stomatitis, *J. Am. Vet. Med. Assoc.*, 129, 95, 1956.
34. **Kuns, M. L.**, The Epizootiology of Vesicular Stomatitis in Middle America, Ph.D. dissertation, University of Wisconsin, Madison, 1962.
35. **Jonkers, A. H., Spence, L., and Aitken, T. H. G.**, Cocal virus epizootiology in Bush Bush forest and the Nariva Swamp, Trinidad, W.I.: further studies, *Am. J. Vet. Res.*, 26, 758, 1965.
36. **Tesh, R. B., Peralta, P. H., and Johnson, K. M.**, Ecologic studies of vesicular stomatitis virus. I. Prevalence of infection among animals and humans living in an area of endemic VSV activity, *Am. J. Epidemiol.*, 90, 255, 1969.
37. **Srinhongse, S.**, Vesicular stomatitis virus infections in Panamanian primates and other vertebrates, *Am. J. Epidemiol.*, 90, 69, 1969.

38. **Trainer, D. O., Glazener, W. E., Hanson, R. P., and Nassif, B. D.,** Infectious disease exposure in a wild turkey population, *Avian Dis.*, 12, 208, 1968.
39. **Donaldson, A. I.,** Bats as possible maintenance hosts for vesicular stomatitis virus, *Am. J. Epidemiol.*, 92, 132, 1970.
40. **Ferris, D., Hanson, R. P., Dicke, R. J., and Roberts, R. H.,** Experimental transmission of vesicular stomatitis virus by diptera, *J. Infect. Dis.*, 96, 184, 1955.
41. **Shelokov, A. I. and Peralta, P. H.,** Vesicular stomatitis virus, Indiana type: an arbovirus infection of tropical sandflies and humans?, *Am. J. Epidemiol.*, 86, 149, 1967.
42. **Sudia, W. D., Fields, B. N., and Calisher, C. H.,** The isolation of vesicular stomatitis virus (Indiana strain) and other viruses from mosquitoes in New Mexico, *Am. J. Epidemiol.*, 86, 598, 1967.
43. **Suarez, O. M.,** Multiplication in and transmission by *Aedes aegypti* (L.) mosquitoes of VSV, *Acta Cient. Venez.*, 18, 387, 1967.
44. **Tesh, R. B., Chaniotis, B. H., and Johnson, K. M.,** Vesicular stomatitis virus, Indiana serotype; multiplication in and transmission by experimentally infected phlebotomine sandflies (*Lutzomyia trapidoi*),. *Am. J. Epidemiol.*, 93, 491, 1971.
45. **Tesh, R. B., Chaniotis, B. H., and Johnson, K. M.,** Vesicular stomatitis virus Indiana serotype: transovarial transmission by phlebotomine sandflies, *Science*, 175, 1477, 1972.
46. **Scherer, W. F., Ed.,** Arthropod-borne Virus Information Exchange, Survey and Recommendation Reports of the Sub-Committee on Arbovirus Laboratory Safety, Am. Comm. on Arthropod-Borne Viruses, 1978.
47. **Berge, T. O., Ed.,** International Catalogue of Arboviruses Including Certain Other Viruses of Vertebrates, Publ. No. (CDC) 75/8301, U.S. Department of Health, Education and Welfare, 1975.
48. **Bergold, G. H. and Munz, K.,** Characterization of Piry virus, *Arch. Gesamte Virusforsch.*, 31, 152, 1970.
49. **Matsumoto, S.,** Electron microscopy of nerve cells infected with street rabies virus, *Virology*, 17, 198, 1962.
50. **Pinteric, L. and Fenje, P.,** Electron microscopic observations of the rabies virus, in *Int. Symp. Rabies, Symposia Series in Immunobiological Standardization*, Vol. 1, S. Karger, Basel, 1966, 9.
51. **Lépine, P., Atanasiu, P.,and Sisman, J.,** Etudes sur le virion rabique, in *Int. Symp. Rabies, Symposia Series in Immunobiological Standardization*, Vol. 1, S. Karger, Basel, 1966, 31.
52. **Hummeler, K., Koprowski, H., and Wiktor, T. J.,** Structure and development of rabies virus in tissue culture, *J. Virol.*, 1, 152, 1967.
53. **Miller, A. and Nathanson, N.,** Rabies: recent advances in pathogenesis and control, *Ann. Neurol.*, 2, 511, 1977.
54. **Warrell, D. A.,** The clinical picture in man, *Trans. R. Soc. Trop. Med. Hyg.*, 70, 188, 1976.
55. **Haig, D. A.,** Rabies in animals, in *Rabies: The Facts*, Kaplan, C., Ed., Oxford University Press, Oxford, 1977, 53.
56. **Debbie, J. G.,** Rabies, in *Progress in Medical Virology*, Vol. 18, Hotchin, J., Ed., S. Karger, Basel, 1974, 241.
57. **Baer, G. M.,** Pathogenesis to the central nervous system, in *The Natural History of Rabies*, Vol. 1, Baer, G. M., Ed., Academic Press, London, 1975, 181.
58. **Murphy, F. A., Bauer, S. P., Harrison, A. K., and Winn, W. C.,** Comparative pathogenesis of rabies and rabies-like viruses. Viral infection and transit from inoculation site to the central nervous system, *Lab. Invest.*, 28, 361, 1973.
59. **Murphy, F. A., Harrison, A. K., Winn, W. C., and Bauer, S. P.,** Comparative pathogenesis of rabies and rabies-like viruses. Infection of the central nervous system and centrifugal spread of virus to peripheral tissues, *Lab. Invest.*, 29, 1, 1973.
60. **Schneider, L. G.,** Spread of virus within the central nervous system, in *The Natural History of Rabies*, Vol. 1, Baer, G. M., Ed., Academic Press, London, 1975, 199.
61. **Schneider, L. G.,** Spread of virus from the central nervous system, in *The Natural History of Rabies*, Vol. 1, Baer, G. M., Ed., Academic Press, London, 1975, 273.
62. **Shope, R. E.,** Rabies, in *Viral Infections of Humans. Epidemiology and Control*, Evans, A. S., Ed., John Wiley & Sons, Chichester, 1976, 351.
63. **Baer, G. M. and Cleary, W. F.,** A model in mice for the pathogenesis and treatment of rabies, *J. Infect. Dis.*, 125, 520, 1972.
64. **Hattwick, M. A. W., Weis, T. T., Stechschultze, C. J., Baer, G. M., and Gregg, M.,** Recovery from rabies. A case report, *Ann. Intern. Med.*, 76, 931, 1972.
65. **Porras, C., Barboza, J. J., Fuenzalida, E., Adaros, H. L., Oviedo de Diaz, A. M., and Furst, J.,** Recovery from rabies in man, *Ann. Intern. Med.*, 85, 44, 1976.
66. **Chalmers, A. W. and Scott, G. R.,** Ecology of rabies, *Trop. Anim. Health Prod.*, 1, 33, 1969.
67. **Bell, J. F.,** Latency and abortive rabies, in *The Natural History of Rabies*, Vol. 1, Baer, G. M., Ed., Academic Press, London, 1975, 331.

67a. **Afshar, A. and Bachmanyar, M.**, Non-fatal rabies virus infections, *Vet. Bull.*, 48, 553, 1978.
68. **Baer, G. M.**, Rabies in nonhematophagous bats, in *The Natural History of Rabies*, Vol. 2, Baer, G. M., Ed., Academic Press, London, 1975, 79.
69. **Sulkin, S. E. and Allen, R.**, Experimental rabies virus infection in bats, in *The Natural History of Rabies*, Vol. 2, Baer, G. M., Ed., Academic Press, London, 1975, 99.
70. **Baer, G. M.**, Bovine paralytic rabies and rabies in the vampire bat, in *The Natural History of Rabies*, Vol. 2, Baer, G. M., Ed., Academic Press, London, 1975, 155.
71. **Constantine, D. G.**, Rabies Transmission by Air in Bat Caves, Public Health Service Publ. No. 1617, National Communicable Disease Center, Atlanta, 1967.
72. **Winkler, W. G.**, Airborne rabies, in *The Natural History of Rabies*, Vol. 2, Baer, G. M., Ed., Academic Press, London, 1975, 115.
73. **Winkler, W. G., Fashinell, T. R., Leggingwell, L., Havard, P., and Conomy, J. P.**, Airborne rabies transmission in a laboratory worker, *J. Am. Med. Assoc.*, 226, 1219, 1973.
74. Rabies in a laboratory worker, *Morbid. Mortal. Weekly Rec.*, 26, 183, 1977.
75. **Hattwick, M. A. W. and Gregg, M. B.**, The disease in man, in *The Natural History of Rabies*, Vol. 2, Baer, G. M., Ed., Academic Press, London, 1975, 281.
76. **Negri, A.**, Beitrag zum Studium der Aetiologie der Tollwuth, *Z. Hyg. Infektionskr.*, 43, 507, 1903.
77. **Derakhshan, I., Bahmanyar, M., Noorsalehi, S., Fayaz, A., Mohammad, M., and Ahouraii, P.**, Light-microscopical diagnosis of rabies. A reappraisal, *Lancet*, 1, 302, 1978.
78. **Kaplan, M. M. and Koprowski, H., Eds.**, *Laboratory Techniques in Rabies*, 3rd ed., World Health Organization, Geneva, 1973.
79. **Johnson, H. N.**, Rabies, in *Viral and Rickettsial Infections of Man*, 3rd ed., Rivers, T. M. and Horsfall, F. L., Jr., Eds., Pitman Medical Publishing, London, and J.P. Lippincott, Philadelphia, 1959, 405.
80. **Steele, J. H.**, History of rabies, in *The Natural History of Rabies*, Vol. 1, Baer, G. M., Ed., Academic Press, London, 1975, 1.
81. **Steck, F.**, Epizootiology of rabies, in *Int. Symp. Rabies (II), Symposia Series in Immunobiological Standardization*, Vol. 21, S. Karger, Basel, 1974, 20.
82. **Taylor, D.**, Rabies: epizootic aspects, *Vet. Rec.*, 99, 157, 1976.
83. **Sodja, I., Lim, D., and Matouch, O.**, Isolation of rabies-like virus from murine rodents, *J. Hyg. Epidemiol. Microbiol. Immunol.*, 15, 229, 1971.
84. **Forster, U., Wachendorfer, G., and Krekel, H.**, Untersuchungen zum Nachweis von Tollwutvirus bei Nagern und Insektivoren-ein Beitrag zur Epidemiologie der Tollwut, *Berl. Muench. Tieraerztl. Wochenschr.*, 17, 335, 1977.
85. **Winkler, W. Cr.**, Fox rabies, in *The Natural History of Rabies*, Vol. 2, Baer, G. M., Ed., Academic Press, London, 1975, 3.
86. **Parker, R. L.**, Rabies in skunks, in *The Natural History of Rabies*, Vol. 2, Baer, G. M., Ed., Academic Press, London, 1975, 41.
87. **McLean, R. G.**, Raccoon rabies, in *The Natural History of Rabies*, Vol. 2, Baer, G. M., Ed., Academic Press, London, 1975, 53.
88. **Allen, R., Sims, R. A., and Sulkin, S. E.**, Studies with cultured brown adipose tissue. I. Persistence of rabies virus in bat brown fat, *Am. J. Hyg.*, 80, 11, 1964.
89. **Allen, R., Sims, R. A., and Sulkin, S. E.**, Studies with cultured brown adipose tissue. II. Influence of low temperature on rabies virus infection in bat brown fat, *Am. J. Hyg.*, 80, 25, 1964.
90. **Sulkin, S. E. and Allen, R.**, Virus infections in bats, *Monogr. Virol.*, 8, 1, 1974.
91. **Baer, G. M., Ed.**, *The Natural History of Rabies*, Vol. 2, Academic Press, London, 1975, ch. 5—7, 10, 14.
92. **Bisseru, B.**, *Rabies*, Heinemann, London, 1972.
93. **Turner, G. S.**, A review of the world epidemiology of rabies, *Trans. R. Soc. Trop. Med. Hyg.*, 70, 196, 1976.
94. **Smith, P. C., Lawhaswasdi, K., Vick, W. E., and Stanton, J. S.**, Enzootic rabies in rodents in Thailand, *Nature (London)*, 217, 954, 1968.
95. Second Conference on the Surveillance and Control of Rabies, World Health Organization, Frankfurt-am-Main, 1977.
96. **Irvin, A. D.**, The epidemiology of wildlife rabies, *Vet. Rec.*, 87, 333, 1970.
97. **Tierkel, E. S.**, Historical review of rabies in Asia, in *Rabies*, Nagano, Y. and Davenport, F. M., Eds., University Park Press, Baltimore, 1971, 3.
98. **Théoridès, J.**, Some remarks on the history of rabies, in *Proc. 23rd Int. Congr. History Med.*, Vol. 2, Wellcome Institute of the History of Medicine, London, 1974, 1252.
99. *Report of the Committee of Inquiry on Rabies*. Her Majesty's Stationery Office, London, 1971.
100. **Zinke, G.**, *Neue Ansichten der Hundswuth, ihrer Ursachen und Folgen, nebst einer sichern Behandlungsart der von tollen Thieren gebissenen Menschen*, Gabler, Jena, 16, 212, 1804.

101. **Galtier, V.**, Physiologie pathologique - les injections de virus rabique dans le torrent circulatoire ne provoquent pas l'éclosion de la rage et semblent conférer l'immunité. La rage peut être transmise par l'ingestion de la matière rabique, *C.R. Acad. Sci.*, 93, 284, 1881.
102. **Galtier, V.**, Etudes sur la rage, *C.R. Acad. Sci.*, 89, 444, 1879.
103. **Pasteur, L., Chamberland, C., and Roux, E.**, Nouvelle communication sur la rage, *C.R. Acad. Sci.*, 98, 457, 1884.
104. **Pasteur, L.**, Méthode pour prévenir la rage après morsure, *C.R. Acad. Sci.*, 101, 765, 1885.
105. **Schoop, G.**, Epidemiology of rabies in the western part of Germany, in *Int. Symp. Rabies, Symposia Series in Immunobiological Standardization*, Vol. 1, S. Karger, Basel, 1966, 237.
106. **Meyer, K. F.**, Can man be protected against rabies?, *Bull. WHO*, 10, 845, 1954.
107. **Tierkel, E. S.**, Canine rabies, in *The Natural History of Rabies*, Vol. 2, Baer, G. M., Ed., Academic Press, London, 1975, 123.
108. **Fleming, G.**, *Rabies and Hydrophobia*, Chapman and Hall, London, 1872.
109. **Pawan, J. L.**, Rabies in the vampire bat of Trinidad, with special reference to clinical course and latency of infection, *Ann. Trop. Med.*, 30, 401, 1936.
110. **Johnson, H. N.**, Derriengue: vampire bat rabies in Mexico, *Am. J. Hyg.*, 47, 189, 1948.
111. **Semple, D.**, Scientific Memoirs by Officers of the Medical and Sanitary Departments of the Government of India, No. 44, 1911.
112. **Umeno, S. and Doi, Y.**, A study on the anti-rabic inoculation of dogs and the results of its practical application, *Kitasato Arch. Exp. Med.*, 4, 89, 1921.
113. **Shimada, K.**, The last rabies outbreak in Japan, in *Rabies*, Nagano, Y. and Davenport, F. M., Eds., University Park Press, Baltimore, 1971, 11.
114. **Kaplan, M. M.**, Epidemiology of rabies, *Nature (London)*, 221, 421, 1969.
115. **Nobuto, K.**, The quarantine of dogs imported into Japan, in *Rabies*, Nagano, Y. and Davenport, F. M., Eds., University Park Press, Baltimore, 1971, 267.
116. **Blamire, R. V.**, Veterinary aspects of rabies, *Postgrad. Med. J.*, 49, 547, 1973.
117. **Acha, P. N. and Hubbard, H. B.**, International management of rabies, in *The Natural History of Rabies*, Vol. 2, Baer, G. M., Ed., Academic Press, London, 1975, 267.
118. World Survey of Rabies, Veterinary Public Health Unit, World Health Organization, Geneva, 1959 to present.
119. Rabies Surveillance for the Americas, Pan American Zoonoses Center, Buenos Aires, 1969 to present.
120. World Survey of Rabies for the Year 1975, Veterinary Public Health Unit, World Health Organization, Geneva, 1977.
121. **Abdussalam, M. and Botton, P. C.**, Global aspects of rabies, in *Int. Symp. Rabies (II), Symposium Series in Immunobiological Standardization*, Vol. 21, S. Karger, Basel, 1974, 3.
122. **Johnson, H. N.**, General epizootiology of rabies, in *Rabies*, Nagano, Y. and Davenport, F. M., Eds., University Park Press, Baltimore, 1971, 237.
123. **Linhart, S. B.**, Biology and control of vampire bats, in *The Natural History of Rabies*, Vol. 2, Baer, G. M., Ed., Academic Press, London, 1975, 221.
124. **Lloyd, H. G.**, Wildlife rabies in Europe and the British situation, *Trans. R. Soc. Trop. Med. Hyg.*, 70, 179, 1976.
125. **Toma, B. and Andral, L.**, Epidemiology of fox rabies, *Adv. Virus Res.*, 21, 1, 1977.
126. **Wachendorfer, G.**, Zur Bekämpfung der Tollwut beim Fuchs, *Prakt. Tierarzt*, 5, 362, 1978.
127. **Kauker, E.**, Vorkommen und Verbreitung der Tollwut in Europa von 1966—74, *Sitzungsber. Heidelb. Akad. Wiss.*, 2, 33, 1975.
128. **Gardner, S.**, personal communication.
129. **Corey, L. and Hattwick, M. A. W.**, Treatment of persons exposed to rabies, *JAMA*, 232, 272, 1975.
130. **Kappus, K. D.**, Canine rabies in the United States, 1971—73: study of reported cases with reference to vaccine history, *Am. J. Epidemiol.*, 103, 242, 1976.
131. **Schwabe, C. W.**, in *Report of the WHO Seminar on Veterinary Public Health*, World Health Organization, Mukteswar, India, 1971, 16.
132. **Abdussalam, M. and Bogel, K.**, The problem of antirabies vaccination, in *Int. Conf. Application of Vaccines Against Viral, Rickettsial and Bacterial Diseases of Man*, No. 226, Pan American Health Organization, World Health Organization, Washington, D.C., 1971, 54.
133. **Bell, J. F.**, Present Concepts of the Epidemiology of Rabies, in *1st Int. Conf. Vaccines*, Sci. Publ. No. 147, Pan American Health Organization, Washington, D.C., 1967, 481.
134. **Bouffard, G.**, Sur l'existence de la rage canine dans le Haut Sénégal et le Niger, *Ann. Inst. Pasteur*, 26, 727, 1912.
135. **Crandell, R. A.**, Arctic fox rabies, in *The Natural History of Rabies*, Vol. 2, Baer, G. M., Ed., Academic Press, London, 1975, 23.

136. **Kemp, G. E., Causey, O. R., Moore, D. L., Odelola, A., and Fabiyi, A.,** Mokola virus. Further studies on IbAn 27377, a new rabies related etiologic agent of zoonosis in Nigeria, *Am. J. Trop. Med. Hyg.*, 21, 356, 1972.
137. **Familusi, J. B. and Moore, D. L.,** Isolation of a rabies related virus from the cerebrospinal fluid of a child with "aseptic meningitis", *Afr. J. Med. Sci.*, 3, 93, 1972.
138. **Familusi, J. B., Osunkoya B. O., Moore, D. L., Kemp, G. E., and Fabiyi, A.,** A fatal human infection with Mokola virus, *Am. J. Trop. Med. Hyg.*, 21, 959, 1972.
139. **Porterfield, J. S., Hill, D. H., and Morris, A. D.,** Isolation of a virus from the brain of a horse with "Staggers" in Nigeria, *Bri. Vet. J.*, 114, 1, 1958.
140. **Shope, R. E.,** personal communication.
141. **Shope, R. E. and Tignor, G. H.,** Rabies and serologically related viruses from Africa, in *Rabies*, Nagano, Y. and Davenport, F. M., Eds., University Park Press, Baltimore, 1971, 53.
142. **Schneider, L. G.,** personal communication.
143. **Murphy, F. A.,** personal communication.
144. **Johnson, H. N.,** Diseases derived from wildlife, *Calif. Health*, 23, 35, 1965.
145. **Whitney, E.,** Flanders strain, an arbovirus newly isolated from mosquitoes and birds of New York State, *Am. J. Trop. Med. Hyg.*, 13, 123, 1964.
146. **Boyd, K. R.,** Serological comparisons among Hart Park virus and strains of Flanders virus, *Infect. Immun.*, 5, 933, 1972.
147. **Metselaar, D., Williams, M. C., Simpson, D. I. H., West, R., and Mutere, F. A.,** Mount Elgon bat virus: a hitherto undescribed virus from *Rhinolophus hildebrandtii eloquens (K. Anderson), Arch. Gesamte Virusforsch.*, 26, 183, 1969.
148. **de Haas, R. A., Jonkers, A. H., and Heinemann, D. W.,** Kwatta virus, a new agent isolated from *Culex* mosquitoes in Surinam, *Am. J. Trop. Med. Hyg.*, 15, 954, 1966.
149. **Kokernot, R. H., McIntosh, B. M., Brooke Worth, C., and de Sousa J.** Isolation of viruses from mosquitoes collected at Lumbo, Mozambique, II. Mossuril virus, a new virus isolated from the *Culex (Culex) sitiens* Wiedemann group, *Am. J. Trop. Med. Hyg.*, 11, 683, 1962.
150. **Teninges, D., Contamine, D., and Brun, G.,** Drosophila sigma virus, *Rhabdoviruses*, Vol. III, Bishop, H. C. D., Ed., CRC Press, West Palm Beach, to be published, chap. 6.
151. **Sano, T.,** Viral diseases of cultured fishes in Japan, *Fish Pathol.*, 10, 221, 1976.
152. **Ahne, W.,** A rhabdovirus isolated from grass carp *(Ctenopharyngodon idella Val.), Arch Virol.*, 48, 181, 1975.
153. **Jahromi, S. S.,** Occurrence of rhabdovirus-like particles in the blue-crab *Callinectes sapidus, J. Gen. Virol.*, 36, 485, 1977.
154. **Jonkers, A. H.,** The epizootiology of the vesicular stomatitis viruses: a reappraisal, *Am. J. Epidemiol.*, 86, 286, 1967.
155. **Crick, J.,** unpublished data.
156. **Schneider, L. G. and Schoop, U.,** Pathogenesis of rabies and rabies-like viruses, *Ann. Inst. Pasteur*, 123, 469, 1972.
157. **Tignor, G. H. and Shope, R. E.,** Vaccination and challenge of mice with viruses of the rabies serogroup, *J. Infect. Dis.*, 125, 322, 1972.
158. **Crick, J.,** unpublished data.
159. **Bauer, S. P. and Murphy, F. A.,** Relationship of two arthropod-borne rhabdoviruses (kotonkan and Obodhiang) to the rabies serogroup, *Infect. Immun.*, 12, 1157, 1975.
160. **Buckley, S. M.,** Arbovirus infection of vertebrate and insect cell cultures with special emphasis on Mokola, Obodhiang and kotonkan viruses of the rabies serogroup, *Ann. N.Y. Acad. Sci.*, 226, 241, 1975.
161. **Murphy, F. A., Johnson, H. N., Harrison, A. K., and Shope, R. E.,** Ultrastructural characterization of a new member of the rhabdovirus group-Klamath virus, *Arch. Gesamte Virusforsch.*, 37, 323, 1972.
162. **Hill, B. J., Underwood, B. O., Smale, C. J., and Brown, F.,** Physico-chemical and serological characterization of five rhabdoviruses infecting fish, *J. Gen. Virol.*, 27, 369, 1975.
163. **Taylor, D.,** Rabies: epizootic aspects, *Vet. Rec.*, 99, 157, 1976.

Chapter 2

SYSTEMS FOR ASSAY AND GROWTH OF RHABDOVIRUSES

H Fred Clark

TABLE OF CONTENTS

I. INTRODUCTION

Rhabdoviruses have been recognized as causing infections in all major orders of vertebrates except the amphibia, and in arthropods and plants as well (see Chapter 1, this volume). These diverse viruses differ remarkably in their ease of experimental cultivation. At one extreme, vesicular stomatitis virus (VSV) may be propagated in many animal systems or in cell culture systems of extraordinarily diverse origins.[1] The ease with which VSV may be grown rapidly to very high titer has led to its development as one of the primary model systems for the study of viral replication. Like VSV, the fish viruses of spring viremia of carp (SVC) and of pike fry (PFV) replicate to high titer with severe cytopathic effect (CPE) in many cell culture systems of differing phylogenetic origin. It is possibly of more than anecdotal interest that these three viruses stand out as possessing virion-bound transcriptase of high activity.[2-4] Rabies virus has been grown in innumerable species of warm-blooded vertebrates and in a great variety of cell culture systems,[5] but usually requires adaptation to cell culture and is irregularly cytopathic.

Other rhabdoviruses have generally been shown to replicate in a limited number of host animal species or cell culture systems. The available broad spectra of rhabdoviruses and also of host cell culture systems present special opportunities for studying virus-host interactions. Examples of the exploitation of this potential may be seen in the numerous published studies of the effect of host upon the pattern of defective virus formation by VSV[6-9] and in the demonstration of phenotypic mixing between VSV and various mammalian or avian oncoviruses.[10-12]

II. *VESICULOVIRUS* GENUS

A. Vesicular Stomatitis Virus

1. Propagation

a. Animals

As in the case of rabies virus (and many other viruses) the systems used for propagation have progressed historically from use of the whole animal to the embryonated egg to cell cultures. The majority of comparative studies have been performed with the New Jersey serotype of VSV. In early reports the Indiana and New Jersey viruses were not distinguished; subsequent studies with identified serotypes have revealed few differences in behavior in vivo.

The apparently first successful experimental infections with VSV were accomplished by inoculation of vesicular fluids from natural equine disease outbreaks onto the scarified tongue of the horse, first reported in France in 1916[13] and subsequently in Kansas in 1917.[14] Although the original isolates were obtained from infected horses, this method of isolating virus was especially useful in identifying VSV outbreaks in cattle or swine, as the potential for infecting horses unequivocally distinguished VSV from foot and mouth disease virus. For many years, serial passage of VSV was accomplished primarily by cultivation in the dorsal tongue epithelium of the horse, or less often the cow.

Approximately a decade later, the guinea pig was determined to be a sensitive host for propagation of VSV.[15,16] Vesicles were induced following inoculation of scarified footpads within 18 to 48 hr after inoculation. More than 700 serial passages of VSV of both Indiana and New Jersey serotypes on scarified guinea pig footpads were reported.[17] Guinea pigs inoculated by the intraperitoneal (IP) route are also infected, developing vesicles on hairless areas of skin approximately one week after infection.

Subsequently VSV was shown to replicate in the intracerebrally (IC) inoculated adult mouse or rat, causing an encephalitis fatal within 48 to 72 hr.[18,19] Newborn and adult mice are equally sensitive to the induction of fatal disease following IC inoculation,[20] but yields of VSV recovered from newborn mouse brain are considerably higher. Mice are susceptible to infection administered by the intranasal, subcutaneous, and oral routes with an efficiency somewhat lower than that obtained by IC inoculation and with increasing resistance noted with aging even after weaning age.[21] At one year of age, mice were found to be completely resistant to infection administered by the intranasal[20] or intramuscular[22] route. Only newborn mice succumb to intraperitoneal inoculation.

Hamsters and chinchillas have also been shown to be highly susceptible to VSV administered by the IC or intranasal route.[23] The virus may also be progagated in the IC-inoculated ferret,[23] rhesus, or cynomolgus monkey,[21] sheep, goat, cow, or horse.[24] Swine develop disease following intravenous or peripheral inoculations.[16,25] Rabbits, cats and dogs are relatively resistant. Newborn chicks inoculated IC do not develop the disease, but the virus can be propagated by inoculation of the scarified tongue of the adult chicken.[26]

VSV may be grown in several species of arthropods. Indiana serotype virus was shown to multiply efficiently in intra-thoracically inoculated *Aedes aegypti* mosquitoes[27] and in phlebotomine flies infected by feeding on viremic hamsters.[28] Both Indiana[29,30] and New Jersey[31] serotypes of VSV have been shown to multiply readily in *Drosophila,* in which they induce sensitivity to a lethal effect from CO_2 exposure similar to that described for sigma virus.

b. Embryonated Eggs

VSV (Indiana and New Jersey) was readily adapted to embryonated eggs inoculated onto the chorio-allantoic membrane (CAM).[32,33] In some studies several passages were required before maximum virus yields were obtained.[33]

Both the age of the embryo and the incubation temperature affect the outcome of the infection.[34] Seven-day-old embryos incubated at 35 to 36° were observed to die in 15 to 16 hr after inoculation, giving virus yields of about 10^8 egg infectious doses (maximum yields are obtained from both CAM and embryo tissue). Older (>10 days) embryos incubated at 39 to 40° often survive the infection; these then exhibit proliferative lesions on the CAM.

c. Cell Culture

Carrel and colleagues reported successful cultivation of VSV in guinea pig bone marrow or minced fetal tissue embedded in plasma clot in 1928.[35] Serial cultivation of both Indiana and New Jersey serotypes in chick embryo tissue in vitro was reported in 1933.[36] Subsequently it was determined that monolayers of trypsin-dispersed guinea pig or calf kidney cells yielded higher titers of virus than did explanted guinea pig tissues or bovine tongue epithelium.[37,38]

VSV is now known to replicate efficiently in almost all known types of mammalian and avian cell culture; particularly high yields are consistently obtained from primary chick embryo[39] and BHK/21 cells. A rapidly cytocidal cytopathic effect (CPE) is usually induced, but induction of polykaryocytosis has been recently described in certain strains of mouse L and C-243-3 cells,[40] BHK/21 cells,[41] and rat XC cells.[42] Cells maintained in suspension culture yield titers of VSV similar to those produced in monolayers.[43,44] The rare mammalian cell types insusceptible to VSV infection include a continuous domestic rabbit kidney cell line DRK3,[45] a rabbit cornea cell line (RC-60),[46] and cultured human polymorphonuclear leukocytes.[47] Cultured human lymphocytes support VSV replication only when stimulated with phytohemagglutinin;[47] murine and human lymphoblastoid cell lines exhibit widely varying degrees of susceptibility to VSV infection.[48]

A majority, but not all, of reptilian and fish-derived cell culture systems have also been found to support replication of VSV to high titer. Reptilian cell lines derived from Grecian tortoise,[49] side-necked turtle,[49] box turtle,[49,50] Tokay gecko,[49] leopard gecko,[51] and Russell's viper[52] as well as primary cultures of Grecian tortoise,[53] spectacled cayman,[49] and reticulated python[49] origin all were highly susceptible to VSV. A green iguana cell line and two box turtle cell lines were found to resist VSV infection.[49] We were also unable to infect the bullfrog cell line FT.[49]

Cells of fish origin also display varying degrees of susceptibility to VSV. The virus replicates readily in rainbow trout (RTG-2), fathead minnow (FHM),[54] and red swordtail (SWT)[55] cell lines. Fish cells found to be resistant include those of a cell line derived from the Hawaiian Omaka[56] and bluegill fry (BF2) and newborn pike fry (NPF) cell lines.[201]

VSV of both Indiana and New Jersey serotypes was shown to replicate without inducing CPE in mosquito cell lines derived from *Aedes albopictus* and *A. aegypti.*[57] Indiana serotype VSV has also been shown to cause noncytopathic infection of a moth-derived cell line.[58] *Drosophila* cells have been shown to support chronic persistent infection with VSV.[59]

The replication of VSV is extremely rapid in highly permissive cell types. Single step growth curves described for suspended chick embryo and monkey kidney cells,[60] and for chick embryo cells[61,62] and mouse L cells[63] in monolayer are essentially identical. New released virus is detectable within 2 hr after the initiation of infection. The virus yield then increases rapidly at a logarythmic rate for 6 to 8 hr; final maximal yields of up to 1000 pfu/cell are obtained by 10 to 12 hr postinfection.

VSV replication occurs in vitro over a very broad range of incubation temperatures. Virus production is near-maximal in mammalian and avian cell cultures at temperatures well above 40°; high yields ($>10^{10.0}$pfu/mℓ) of virus have also been obtained following prolonged incubation of snake (*Vipera russelli*) cell cultures at temperatures as low as 11°.[201] Decrease in virus yield occurs following serial passage of VSV at a high multiplicity of infection, a phenomenon mediated by defective interfering (DI) particles which mechanism has been extensively studied (see Chapter *11*, Volume *II*). The efficiency of DI particle-mediated interference is affected markedly by the host cell type employed. Madin-Darby bovine kidney cells have been reported to support DI particle formation less well than other mammalian cell types.[6] In a comparison of three reptilian cell lines infected with VSV, the lizard cell line GLI was reported to produce selectively enhanced numbers of DI virions.[8] VSV-induced interferon induction has also been reported to cause reduced yields in cell culture, but only in the case of multiple generation growth necessitated by infection at a low multiplicity of infection.[64]

2. Assay

VSV preparations may be efficiently titrated for their infectious dose content by inoculation of serial virus dilutions onto the scarified tongue of the horse, cow, or rooster or the footpad of the guinea pig, into the 7-day-old embryonated hen's egg by a variety of routes, into IC-inoculated adult or newborn mice, or into cell culture. The relative sensitivities of the different assay systems may differ according to the previous passage history of the test virus, especially with animal-passaged virus[26,65] but maximum sensitivity for most currently utilized cell culture-adapted strains may be obtained by use of cell culture assays, the embryonated egg, or the IC-inoculated mouse.

Maximum precision is obtained by plaque assay in cell culture under solid overlay. Any cell culture system in which VSV is cytopathic may be used, but chick embryo and BHK/21 cells are especially sensitive. An exhaustive study of factors affecting the sensitivity of plaque assay in chick embryo cells has been reported by Valle.[66] Chick embryo fibroblasts were found to give higher plaque titers than nine mammalian cell types. Titers were maximized by using 1-day-old cultures of 6-day-old embryos, by replacing agar with agarose at minimum usable concentration as a solidifying agent, and by incorporating either poly-L-ornithine (30 μg/mℓ) or DEAE dextran (5 mg/mℓ) into the inoculum. A minimum volume of inoculum was also found to give maximum plaquing efficiency.

VSV propagated in cell culture may also be assayed by hemagglutination (HA). Optimal conditions require the use of goose erythrocytes incubated at 4° in the absence of sera which usually contain nonspecific inhibitors. A pH of 5.8 is optimal for the Indiana serotype; pH 6.4 is optimal for New Jersey virus.[67] HA activity is associated with the intact virion.[68]

The complement fixation test has been successfully applied to the assay of intact standard and defective VSV virions and of numerous subcomponents of experimentally disrupted virions. Extensive studies of the various complement-fixing subcomponents of VSV have been recently reviewed by Knudson[69] and by Howatson.[70]

B. Other Vesiculovirus Group Viruses

Cocal, Piry, and Chandipura viruses and the Brazil and Argentine serotypes of VSV have been extensively compared with VSV Indiana and New Jersey serotype viruses using virus stocks prepared by identical methods in BHK/21 cells.[71,72] Each is basically similar to VSV in host range; each may be efficiently propagated in newborn mouse brain and numerous cell culture systems and assayed by plaque technique in cell systems useful for VSV.[72-74] Cocal virus has been reported to be more temperature-sensitive than VSV Indiana and VSV New Jersey viruses in BHK/21 cells: the maximum temperature permissive for optimal plaque formation was found to be 37°.[75] Substrains adapted to higher temperatures have been developed.[76]

Both Cocal and Piry viruses have been demonstrated to grow in mosquitoes[73,74] and in *Aedes aegypti* and *A. albopictus* mosquito cell lines.[74,77] Cocal, Chandipura, and Piry viruses have each been demonstrated to multiply in *Drosophila*, thereby inducing sensitivity to CO_2.[31,78]

III. *LYSSAVIRUS* GENUS

A. Rabies Virus

1. Propagation

Although rabies virus was successfully transmitted to experimental animals as early as 1879,[79] its growth either in vivo or in vitro is often so slow that rabies has sometimes been included among the "slow virus" diseases.[80] Nevertheless, the importance of producing large quantities of viral antigen for vaccine purposes has led to continued and extensive study of propagation substrates.

a. Animals

The earliest experimental transmissions of rabies virus were accomplished primarily by the use of rabbits and dogs in the latter part of the 19th century (see Chapter 4, Volume III). Subsequently, essentially all species of mammals tested have been found to be susceptible, although wide variation in sensitivity occurs. The successful infection of a number of avian species has also been reported as has prolonged persistence (if not increase in titer) of virus in certain species of reptiles. The host range has been thoroughly reviewed by van Rooyen and Rhodes.[81]

Mammals are invariably most susceptible to infection induced via the intracerebral route which infection also is characterized by the shortest and most consistent incubation periods. Viruses "fixed" by adaptation by prolonged serial passage in rodents or lagomorphs commonly cause death in 5 to 7 days after IC incubation. Characteristically, feral or street strains of rabiesvirus are slightly less effective in inducing infection by peripheral routes of inoculation than by IC inoculation; fixed strains show greatly diminished capacity for inducing infection by peripheral routes.[82]

Despite the very long history and widespread experience with propagation of fixed rabies virus in animal brains, there is very little data published on relative yields of virus in different species. Investigators have been interested in immunogenic mass, not virus titers! The cow, sheep, goat, and rabbit have been extensively used for brain and spinal cord vaccines.[81,83-85] In order to avoid the danger of allergic encephalomyelitis associated with adult brain tissue, vaccines have also been prepared in newborn mouse,[86,87] and less often newborn rat[88] and newborn rabbit[89] brain.

Rabies virus propagated in the adult or newborn mouse brain has been most commonly used for laboratory studies, with newborn mice tending to yield higher titers. Habel[90] titrated 31 fixed strains of rabies virus propagated in rabbit brain and found little variation in titer; a concentration of 10^4 mouse lethal doses was most often detected. Adaptation of similar rabies strains to mice was subsequently shown by Mason[91] to lead to production in mice in titers up to 10^6 $MLD_{50}/0.03$ $m\ell$ of brain suspension. Titers of unadapted street virus are commonly lower by a factor of up to 1000-fold.[91,92]

In studies of wild animals infected with street rabies virus, higher yields of virus have sometimes been obtained from salivary glands than from brain tissue. Intramuscularly inoculated skunks and foxes were shown to frequently produce virus in salivary glands in concentrations 10- to 100-fold greater than that found in brain.[93] Titers as high as 10^6 $MLD_{50}/0.03$ $m\ell$ were recovered from skunk saliva,[94] indicating virus present in a concentration approaching levels required for purification and detailed biochemical characterization. The high yield from saliva is explained by electron microscopic studies showing that, in contradistinction to patterns observed in nervous tissue, maturation of virus in salivary glands occurs selectively by a process of uniform budding of complete standard virions from acinar cell plasma membranes.[95]

Special interest has also attended the growth of rabies virus in bats. Constantine[96] demonstrated similar degrees of sensitivity to infection with street virus in eleven species of insectivorous bats; most of these produced a concentration of virus in brain tissue equal to that expected in mice. Assays of tissues of Mexican free-tailed bats (*Tadarida brasiliensis mexicana*) dying from rabies infection induced by varying routes of inoculation revealed consistently highest titers in brain tissue.[97] Contrary results were described in little brown bats (*Myotis lucifugus*): experimentally infected animals were reported to often exhibit virus in brown fat or salivary gland in the absence of brain infection.[98]

b. Embryonated Eggs

Rabies virus was first propagated in embryonated hen's egg by Kligler and Bernkopf[99] who had partial success in passaging infected chorioallantoic membrane (CAM) tissues, but observed maximum virus yields in the chick embryo brains. Subsequent early studies were reviewed by Koprowski and Cox.[100] The latter authors demonstrated that both fixed and street rabies strains could be readily adapted to embryonated eggs provided that 7-day-incubated eggs were inoculated into the yolk sac and that embryo tissues were not harvested until 8 to 12 days postinoculation. Under these conditions, maximum virus concentrations were attained in the chick embryo brain and body, followed by progressively lower titers detected in the CAM, yolk sac, amniotic and allantoic fluids. Following adaptation, virus yields of $10^{4.5}$ to $10^{5.5}$ infectious doses per 0.03 cm of chick embryo tissue were consistently obtained.

Yoshino et al.[101] found the 1-day-old embryonated egg inoculated into the yolk sac to be particularly sensitive in infection with fixed or street strains of rabies virus. Peak titers of $>10^{8.0}$ infectious doses per gram of embryo were obtained 3 days after inoculation of 1-day-old eggs with egg-adapted Nishigahara or Flury HEP viruses.[102]

Rabies virus is now commonly propagated in duck embryos for vaccine production. Several strains of virus have been successfully adapted to duck embryos by serial inoculation of 7-day-old embryos inoculated into the yolk sac or allantoic activity.[103,104] The embryo tissues are harvested 10 to 14 days after inoculation; infected embryos yield as much as $10^{8.5}$ MLD_{50} of virus per gram of tissue.[104]

c. Cell Culture

The efficient propagation of rabies virus in cell culture has only been achieved relatively recently. Nearly all mammalian and avian cell types tested can now be infected with rabies virus, but many systems require prolonged periods of adaption before high yields of virus are produced. The subject has recently been thoroughly reviewed by Wiktor and Clark.[5]

Noguchi[105] and Levaditi[106] early reported prolonged replication of virus in cultured fragments of nervous tissue of infected animals. Later investigations demonstrated multiplication of fixed rabies viruses inoculated onto mouse or rat embryo brain cultures.[107-109] Multiplication of rabies in nonnervous tissue cell culture was subsequently reported with mouse ependymoma,[110] and primary mouse[111] and hamster[112] kidney cell cultures. The chick embryo-adapted Flury HEP strain of rabies virus multiplies readily in primary duck embryo fibroblasts.[113]

Rabies virus for experimental use is now most commonly produced in certain continuous cell lines which yield high titers of fixed virus after brief (or no) periods of adaptation. These include the hamster cell lines BHK/21,[114] NIL/2[5] and CER[202] and more recently neuroblastoma cell lines of murine,[115,116] or human[116] origin. Rabies virus for vaccine use may be grown in a variety of primary cell culture systems (hamster kidney,[112] monkey kidney,[117] dog kidney,[118] pig kidney,[119] etc.), although these usually yield somewhat lower titers than the above-mentioned cell lines. Much of the virus used for human vaccines is now being produced in human diploid cell strains. Adaptation of virus to these cells required prolonged serial passage of infected cells, often with addition of fresh uninfected cells as chronically infected cells lost the capacity to divide.[120] Again, virus adapted to these cells produced a maximum yield of about$10^{7.0}$ infectious doses per mℓ, a yield about tenfold lower than that obtained from highly susceptible cell lines.

Fixed rabies virus strains have also been propagated in cells of poikilothermic vertebrates. Several strains of BHK/21 cell-adapted virus grow in VSW cells, a *Vipera* cell line that spontaneously produces C-type virus,[52] to titers similar to those obtained in BHK/21 cells.[121] Rabies virus was serially propagated in other cell lines derived from snakes, and from the Tokay gecko (*Gekko gecko*),[121] the side-necked turtle (*Podocnemis unifilis*),[121] and the lizard *Anguis fragilis.*[122] Cells of snakes of the families Viperidae and Crotalidae appear to be more susceptible to rabies virus than cells of other reptiles studied to date.[203] Serial cultivation of Flury HEP rabies virus has been described in fathead minnow cells,[54] but we have been unable to demonstrate efficient replication of rabies virus in any of a variety of fish or amphibian cells tested,[121] (unpublished results).

Although numerous authors have described the production of defective interfering rabies virions (see Chapter 9, Volume II), "auto-interference" in cultures infected at high multiplicity is not normally a problem. Extensive studies with BHK/21 cells have indicated that yields of released virus often vary directly with virus input multiplicity.[5] Exceptional instances of reduced yield have been noted in chick embryo fibroblasts inoculated at high input multiplicity[123] and in C1300 mouse neuroblastoma cells in which fixed virus was serially passaged without dilution (unpublished data).

The description of "single-cycle" growth curves in rabies virus-infected cells has not been entirely satisfactory because (as indicated by immunofluorescent studies) cell infection proceeds asynchronously even with very high input multiplicities.[5] However, growth curve studies of fixed rabies virus-infected BHK/21 cells inoculated at virus multiplicities ranging from 5.0 to 200 have similarly indicated the first appearance of released infectious virus at 12 to 15 hr postinfection.[124-126] Cell cultures infected at the highest multiplicity produced a maximum titer of released virus in 24 hr; cells infected with the lowest multiplicity required 72 hr to produce maximum levels of released virus.[126] Electron microscopic examination revealed *de novo* budding of virus from plasma membrane as early as 6 hr following infection at high multiplicity.[124]

Virus yields may be profoundly affected by the nature of the serum protein supplement, by pH, and by incubation temperature. Highly susceptible cells produce maximum yields in media supplemented only with serum albumin; less susceptible cells may require the presence of 2 to 10% serum. Because acidic medium inhibits production of rabies virus, infected cells are normally fed with medium adjusted to a pH of 7.6 to 8.0.[5] Temperature gradient studies[127] have revealed that although virus replication in mammalian cells proceeds over a temperature range of at least from 20 to 42°, maximum accumulation of virus at 3 to 7 days is consistently obtained in cultures incubated at 32 to 33°. A contrary result has been observed in certain partially permissive reptilian cell lines in which maximum virus yields are obtained at 35 to 39° and little or no virus is produced at lower temperatures (unpublished data).

2. Assay

a. Infectivity

For many years the mouse has been the primary host used for isolation of street rabies virus and titration of fixed virus; the mouse is at least a tenfold more sensitive host than the rabbit or guinea pig.[81] The intracerebral route of inoculation is preferred, being from 10- to 100-fold or more sensitive than peripheral routes, depending on virus strain. The intracerebrally inoculated newborn mouse (less than 7 days old) is the most susceptible host known.[128,129] We regard failure to kill newborn mice as the definitive test of lack of rabies infectivity, as all attenuated viruses tested to date have retained an unaltered capacity to lethally infect newborn mice. Incubation periods vary upwards from 3 days for development of paralytic or encephalitic signs and 4 days for death, depending on the age of the mice and the strain of virus.

Little is known about the effect of mouse strain on sensitivity. Use of Swiss mice selected for susceptibility to neurotropic viruses has been recommended,[130] but the relative susceptibility of thoroughly defined inbred strains of mice currently available has not been well studied. Infectivity of rabies virus preparations for adult mice may be enhanced slightly by addition of DEAE dextran to the inoculum[131] or by treatment of mice with corticosteroids.[132] Maintenance of infected mice at high (35°) ambient temperature greatly increases their survival rate,[133] a phenomenon that is also partially virus strain-dependent.[134]

Certain cell lines also provide a sensitive system for isolation of rabies virus. Street virus strains infect BHK/21,[135] CER,[136] and C1300 mouse neuroblastoma cells[137] with an efficiency comparable to mice. Such infection must usually be demonstrated by immunofluorescent staining which may reveal specific antigen as early as 2 days after infection. Quantitative assay may be obtained in these cell systems by counting immunofluorescent foci.[136,138,139]

The most practical precise assay of rabies virus is obtained by plaque assay, a technique currently only applicable to cell culture-adapted fixed strains of virus. The most widely and successfully applied plaque assay technique is based upon the inoculation of agarose-suspended BHK/21/13S cells.[140,141] Most cell culture-adapted fixed rabies viruses produce clear plaques within 5 to 7 days after inoculation in this system. Many fixed rabies viruses also from clear plaques on monolayers of CER cells (Wiktor, unpublished data). Plaque assay of rabies virus has also been described using ERA strain virus adapted to growth on pig kidney cell line PK-2A,[142] and various cell culture adapted rabies virus strains inoculated onto BHK/21 cells overlaid with Sephadex G200® beads.[143] Plaques consisting of proliferative foci rather than cytolysis have been demonstrated by inoculation of certain fixed rabies viruses onto chick embryo fibroblast cell cultures.[144]

b. Staining of Rabies Antigens, Hemagglutination, and Serological Assay

Although animals dying of rabies normally exhibit characteristic incubation periods and clinical signs, confirmation of rabies virus etiology requires the demonstration of either Negri bodies or, more reliably, immunofluorescent antigen (IFA) in brain tissues. Confirmation of infection of cell culture may also require demonstration of rabies IFA.

Negri bodies arz characteristic intra-neuronal cytoplasmic inclusions best visualized by Seller's stain; knowledge of their appearance, composition, and distribution in the brain has recently been reviewed by Atanasin.[145] Staining of rabies IFA is most often performed by the direct technique using acetone-fixed brain impression smears or cell culture preparations.[146] The predominant antigen detected in acetone-fixed cells is viral nucleocapsid; immunofluorescent staining of unfixed cells reveals primarily viral glycoprotein antigen localized in cell plasma membrane.[5]

Rabies virus of several fixed strains has been shown to agglutinate goose erythrocytes at 4°C at an optimum pH of 6.2.[67,125] The hemagglutinating activity, absent in disrupted preparations, is associated with the complete viral particle.[125,147] Hemagglutination requires a virus concentration of at least $10^{6.0}$ plaque-forming unit (PFU)/mℓ. Because the reaction is inhibited by serum lipoproteins,[148] virus to be assayed by hemagglutination must be grown in cell cultures fed with serum-free medium.

Rabies virus and its components may be assayed by the complement-fixation (CF) test. The antigenic composition of the virion was recently reviewed by Aaslestad:[149] present knowledge indicates that the nucleocapsid is the predominant CF antigen. Rabies virus infection of cells may also be indicated by specific complement-dependent immune lysis of infected cells[150] and by radioimmune assay.[151]

B. Other Lyssavirus Group Viruses

Lagos bat and Mokola viruses kill weanling as well as newborn mice inoculated IC; Kotonkan and Obodhiang viruses kill only newborn mice.[152] Lagos bat and Mokola viruses grow efficiently in BHK/21 cells and, following adaptation to these cells, may be assayed by plaque technique in agarose-suspended BHK/21/13S cells.[153] Each of the rabies serogroup viruses can be grown without prior adaptation in CER[136] and mouse C1300 neuroblastoma cells.[201] Kotonkan and Obodhiang viruses are cytopathic for C1300 cells; $TCID_{50}$ titers can be obtained by determining cytopathic effect endpoints following inoculation of serial virus dilutions. Kotankan and Obodhiang viruses replicate in cells of Singh's *Aedes albapictus* mosquito cell line; viruses propagated in the mosquito cells were subsequently found to be adapted to serial passage in BHK/21 and VERO cells.[154]

IV. BOVINE EPHEMERAL FEVER VIRUS (BEFV)

Although the earliest work on this virus was confined to transmission studies in cattle, a number of authors subsequently succeeded in adapting virus from bovine blood or buffy coat to growth in the IC-inoculated newborn mouse.[155-158] Within a few passages in newborn mice the incubation period became stabilized at 2 to 3 days and brain titers up to $10^{8.0}$ $MLD_{50}/m\ell$ were attained. Virus has also been isolated directly in newborn hamsters.[159] Although only very young mice (less than 3 days old) were useful for primary isolation of the virus, suckling mouse brain-adapted virus could be successfully transmitted to adult mice, and to newborn rats and guinea pigs.[160] Larger laboratory and domestic animals (other than cattle) are refractory to experimental infection.[161]

Ten-day-old chick embryos were successfully infected by intravenous inoculation of BEFV.[160] However, only 25% of infected embryos failed to hatch; the remainder hatched into baby chicks with severe nervous signs and anatomical deformities.

Only a limited number of cell culture systems have been reported to be highly sensitive to infection with BEFV. These include BHK/21,[159,161-163] VERO,[162,164] and MS[162] (monkey kidney origin) cell lines. In each of these systems, cultures infected at high multiplicity develop CPE within 24 to 48 hr and produce maximum yields of released virus (approximately $10^{6.0}$ $TCID_{50}/m\ell$ in 48 to 72 hr). A single step growth curve indicating a latent period of 8 to 12 hr has been described in MS cells.[162] BEFV has been most successfully plaque assayed in VERO cells; optimum plaque size in this system is obtained after 7 to 10 days of incubation.[162,165]

V. OTHER MAMMALIAN RHABDOVIRUSES

The preponderance of the remaining known rhabdoviruses of mammalian, avian, or arthropod origin (see Chapter 4, this volume) have been isolated in newborn mice and have been little characterized in other systems. The most thoroughly studied of these is Kern Canyon virus which replicates with no required period of adaptation to high titer in BHK/21 cells.[166,167] Similarly high titers ($10^{7.0}$ to $> 10^{8.0}$ pfu/mℓ) may be obtained by propagation in C1300 mouse neuroblastoma cells, but many other cell systems are refractory to infection (unpublished data). Kern Canyon virus may be titrated by the agarose-suspended BHK/21/13S cell system used for rabies virus.[140]

Kwatta and Navarro viruses have been grown in VERO cells as well as in newborn mouse brain.[168] Klamath virus has been grown in BHK/21 cells, in which CPE is induced.[169] Although Flanders (Hart Park) virus replicates in newborn mouse brain,[170-172] attempts at propagation in a variety of cultured cells were disappointing.[170] We have recently achieved serial propagation of this virus in C1300 cells with destructive CPE enabling ready determination of $TCID_{50}$ endpoints. Released virus titers at early passage levels are ca. $10^{5.0}$ $TCID_{50}/m\ell$; similar titers have been achieved in human neuroblastoma cell line IMR32.[201] It would appear that attempts at propagation in neuroblastoma cells are highly justified for the numerous neurotropic rhabdoviruses not yet cultivated in cell culture.

VI. FISH RHABDOVIRUSES

The four known rhabdoviruses of fish have each been propagated efficiently in cell cultures maintained at submammalian temperatures, the Egtved virus and infectious hematopoietic necrosis virus (IHNV) of trout selectively in fish cell cultures and the spring viremia of carp (SVC) and pike fry (PF) viruses in mammalian cells as well. The ease of cultivating and assaying the latter two viruses has led to their status amongst the best characterized of the rhabdoviruses.

Egtved virus of European rainbow trout was shown to be a filterable infectious agent by fish transmission studies in 1954.[173] While primarily a natural infection of rainbow trout, Egtved virus may also be artificially transmitted to brown trout, brook trout, and Baltic salmon.[174] The incubation period varies from about 7 days to 3 weeks for parenterally inoculated fish;[174,175] low water temperatures (<8.0°) increase the duration and lethality of disease.[175]

Egtved virus was first cultivated in vitro in primary rainbow trout ovary cell culture.[176] It was subsequently also cultivated in primary gonadal cultures of the tench and carp,[177] and in the cell lines rainbow trout gonad (RTG-2),[178,179] and fathead minnow (FHM).[179] In FHM cells, the optimum temperature for viral replication was determined to be 14°. Comparative single step growth curves have been described in both FHM and RTG-2 cells. The virus latent periods were 8 and 12 hr in FHM and RTG-2 cells respectively. Peak released virus titers of $10^{7.5}$ pfu/mℓ were obtained in 20 (FHM) to 48 (RTG-2) hr.[179] Yields of Egtved are decreased if medium is supplemented with bovine sera albumin rather than whole serum.[180]

Plaque assay of Egtved virus has been described in agar or agarose-overlaid FHM and RTG-2 cell cultures incubated at 14 to 15°.[179,180] Clear plaques require 4 (FHM) to 6 (RTG 2) days for development. Detection of Egtved viral antigen in infected fish tissues and cell cultures using the direct fluorescent antibody technique employing hyperimmune serum produced in rabbits has also been described.[181]

Three originally separately described rhabdovirus isolates of North America salmonid fish have now been collected into the species infectious hematopoietic necrosis virus.[182] The virus may be experimentally transmitted to fingerling salmonid fish with a sensitivity partially related to the host of origin of the strain.[182-184] Water temperature also affects the outcome of experimental infection; elevation to temperatures to $\geq$ 13° shortly after infection prevents disease.[185]

IHNV was first cultivated in cell culture in primary sockeye salmon embryo cell cultures; titers of $10^{6.5}$ to $10^{7.0}$ $TCID_{50}$/0.2 mℓ were obtained within 2 to 6 days of incubation at temperatures ranging from 10 to 20°.[186] Similar yields were subsequently reported in cell cultures of the salmonid fish cell lines STE-137 or CHSE-214[187] and in FHM cells.[188] Because of autoinference presumably mediated by defective interfering virions, maximum yields of IHNV are attained only if serial passage is performed with diluted inocula.[189]

A single step growth cycle of IHNV in CHSE-214 cells at 18° revealed appearance of new virus within 5 hr, a maximum titer of cell-associated virus at 20 hr, and maximum released virus concentration at 48 hr postinfection.[190] IHNV plaque assay may be performed in CHSE-214 cells overlaid with agar and incubated for 6 days at 18° prior to observation.[191]

Spring viremia of carp virus (SVCV) causes acute disease in carp inoculated via a variety of routes. Fish are much more sensitive to parenteral (especially swim bladder) inoculation than to oral infection.[192] Virus was isolated first in primary carp ovary tissue and subsequently in FHM cells.[193] Fish cell lines BF-2 and BB are susceptible but less sensitive than FHM cells. Maximum virus sensitivity was obtained in cell cultures incubated at 20 to 22°, compared with an optimum temperature of 16 to 17° in vivo.[193] A single step growth curve of SVCV in FHM cells indicated appearance of new virus within 5 hr and a maximum yield of over $10^{8.0}$ pfu/mℓ attained by 10 hr postinfection.[194] Plaque assay of the virus may be performed in FHM cells; plaques appear within 2 to 3 days.[194,195]

SVCV as also been shown to replicate efficiently in nonpiscine cells including chick embryo fibroblasts, mammalian BHK/21, VERO, WI-38, HEP-2, and MDCK cells and reptilian VSW, VH2, GLI and TH-1 cells.[195] The virus was cytopathic in these systems at 23°; plaque assays were performed conveniently in BHK/21 cells.[1] Egtved virus replicated in many of the same cell systems at 18°, but, unlike the case with SVCV, released virus yields were consistently below those obtained in trout (RTG-2) cells.[1]

The fourth rhabdovirus of fish origin, pike fry virus (PFV), has been successfully transmitted to 5 to 6 cm long pike fry in which it induces lethal disease[196] and to FHM cells in which it replicates rapidly to high titer.[196,197] Maximum titers were obtained only when serial passages were performed at a low multiplicity of infection. Single cycle growth curves indicated optimum replication at 21 to 28° in FHM cells, with an eclipse period of about 4 hr and maximum yields of released virus of up to $10^{9.5}$ pfu/mℓ produced by 12 hr.[197] Plaque assay may be performed in FHM[197] or BHK/21 (unpublished data) cells incubated for 2 days at 23°.

PFV replicates vary efficiently in BHK/21 cells[4] as well as in human (WI-38) fibroblasts (unpublished data). Maximum released virus yields in excess of $10^{9.0}$ pfu/mℓ were obtained in PFV-infected BHK/21 cells incubated over a temperature range of 22 to 28°. BHK/21 cells infected with PFV (or SVCV) and incubated at 20 to 22° may provide a uniquely useful system for the study of viral synthetic processes in cells metabolically quiescent because of low incubation temperatures rather than metabolic poisons.[1] PFV and SVCV have also been shown to replicate in vivo in *Drosophila* inducing CO_2 sensitivity; IHNV and Egtved viruses did not infect *Drosophila.*[198]

Standard cell culture systems have been suggested for simply detecting any of the four fish rhabdoviruses. Wolf[199] suggested FHM cells incubated at 20° coupled with RTG-2 cells incubated at 15°, both in a tris-$NaHCO_3$ buffered medium in equilibrium with the atmosphere, as an ideal method of detecting fish rhabdoviruses. Hill et al.[200] described plaque assay of all four viruses in epithelioma papulosum carpio cells under agarose overlay; the trout viruses were incubated at 14° and SVCV and PFV at 20°.

VII. REPTILE RHABDOVIRUSES

Propagation of reptilian-origin Marco, Chaco, and Timbo viruses is discussed in Chapter 10, Volume III. Each virus causes lethal disease in IC-inoculated newborn mice or hamsters. Each virus may be propagated in BHK/21 or VERO cells incubated at 30°. Chaco and Marco (but not Timbo) viruses also replicate in a variety of reptilian and amphibian cell lines at 30°, but these cell systems were not found to be more sensitive than BHK/21 cells.

REFERENCES

1. **Clark, H F. and Soriano, E. Z.,** Fish rhabdovirus replication in non-piscine cell culture: New system for the study of rhabdovirus-cell interaction in which the virus and cell have different temperature optima, *Infect. Immun.*, 10, 180, 1974.
2. **Chang, S. H., Hefti, E., Obijeski, J. F., and Bishop, D. H. L.,** RNA transcription by the virion polymerases of five rhabdoviruses, *J. Virol.*, 13, 652, 1974.

3. **Roy, P. and Clewley, J. P.**, Spring viremia of carp virus RNA and virion-associated transcriptase activity, *J. Virol.*, 25, 912, 1978.
4. **Roy, P., Clark, H F., Madore, H. P., and Bishop, D. H. L.**, RNA polymerase associated with virions of pike fry rhabdovirus, *J. Virol.*, 15, 338, 1975.
5. **Wiktor, T. J. and Clark, H F.**, Growth of rabies virus in cell culture, in *The Natural History of Rabies*, Vol. 1, Baer, G. M., Ed., Academic Press, New York, 1975, chap. 9.
6. **Huang, A. S. and Baltimore, D.**, Defective viral particles and viral disease processes, *Nature (London)*, 226, 325, 1970.
7. **Perrault, J. and Holland, J.**, Variability of vesicular stomatitis virus autoinferference with different host cells and virus serotypes, *Virology*, 50, 148, 1972.
8. **Lunger, P. D. and Clark, H F.**, Host effect on vesicular stomatitis virus morphogenesis and "T" particle formation in reptilian, avian, and mammalian cell lines, *In Vitro*, 11, 239, 1975.
9. **Potter, K. N. and Stewart, R. B.**, Comparison of vesicular stomatitis virus defective interfering particle synthesis in chick embryo and L cells, *Can. J. Microbiol.*, 22, 1458, 1976.
10. **Zavada, J.**, Pseudotypes of vesicular stomatitis virus with the coat of murine leukaemia and of avian myeloblastosis viruses, *J. Gen. Virol.*, 15, 183, 1972.
11. **Weiss, R. A., Boettiger, D., and Love, D. N.**, Phenotypic mixing between vesicular stomatitis virus and avian RNA tumor viruses, in Cold Spring Harbor Symp. Quant. Biol., 39, 913, 1974.
12. **Huang, A. S., Besmer, P., Chu, L., and Baltimore, D.**, Growth of pseudotypes of vesicular stomatitis virus with N-tropic murine leukemia virus coats in cells resistant to N-tropic viruses, *J. Virol.*, 12, 659, 1973.
13. **Vigel, M.**, Sur le stomatite érosive, *Bull. Soc. Cent. de Med. Vet.*, 69, 59, 1916.
14. **Eichorn, A.**, Vesicular stomatitis in cattle, *Am. J. Vet. Med.*, 12, 162, 1917.
15. **Cotton, W. E.**, Vesicular stomatitis in its relation to the diagnosis of foot-and-mouth disease, *J. Am. Vet. Med. Assoc.*, 69, 313, 1926.
16. **Olitsky, P. K., Traum, J., and Schoening, H. W.**, Comparative studies on vesicular stomatitis and foot-and-mouth disease, *J. Am. Vet. Med. Assoc.*, 70, 147, 1927.
17. **Wagener, K.**, Infection and immunity of vesicular stomatitis in guinea pigs, *Vet. Met. Kansas City M.*, 26, 388, 1931.
18. **Syverton, J. T., Cox, H. R., and Olitsky, P. K.**, Relationship of the viruses of vesicular stomatitis and of equine encephalomyelitis, *Science*, 78, 216, 1933.
19. **Cox, H. R. and Olitsky, P. K.**, Further studies on neurotropism of vesicular stomatitis virus, *Proc. Soc. Exp. Biol. Med.*, 30, 654, 1933.
20. **Sabin, A. B. and Olitsky, P. K.**, Influence of host factors on neuroinvasiveness of vesicular stomatitis virus. I. Effect of age on the invasion of the brain by virus instilled in the nose, *J. Exp. Med.*, 66, 15, 1937.
21. **Olitsky, P. K., Cox, H. R., and Syverton, J. T.**, Comparative studies on the viruses of vesicular stomatitis and equine encephalomyelitis, *J. Exp. Med.*, 59, 159, 1934.
22. **Sabin, A. B. and Olitsky, P. K.**, Influence of host factors on neuroinvasiveness of vesicular stomatitis virus. II. Effect of age on the invasion of the peripheral and central nervous systems by virus injected into the leg muscles or the eye, *J. Exp. Med.*, 66, 35, 1937.
23. **Kowalczyk, T. and Brandly, C. A.**, Experimental infections of dogs, ferrets, chinchillas, and hamsters with vesicular stomatitis virus, *Am. J. Vet. Res.*, 15, 98, 1954.
24. **Frank, A. H., Appleby, A., and Siebold, H. R.**, Experimental intracerebral infection of horses, cattle, and sheep with the virus of vesicular stomatitis, *Am. J. Vet. Res.*, 6, 28, 1945.
25. **Mohler, J. R.**, Vesicular stomatitis of horses and cattle, *J. Am. Vet. Med. Assoc.*, 52, 410, 1917.
26. **Fellowes, O. N., Dimopoulos, G. T., Tessler, J., Hess, W. R., Vardsman, T. H., and Callis, J. J.**, Comparative titrations of vesicular stomatitis virus in various animal species and in tissue culture, *Am. J. Vet. Res.*, 17, 799, 1956.
27. **Mussgay, M. and Suarez, O.**, Multiplication of vesicular stomatitis virus in *Aedis aegypti* (L.) mosquitoes, *Virology*, 17, 202, 1962.
28. **Tesh, R. B., Chaniotis, B. N., and Johnson, K. M.**, Vesicular stomatitis virus, Indiana serotype: Multiplication in and transmission by experimentally infected phlebotomine sandflies (*Lutzomyia trapidoi*), *Am. J. Epidemiol.*, 93, 491, 1971.
29. **Périès, J., Printz, P., Canivet, M., Chuat, J. -C.**, Multiplication du virus de la stomatite vésiculaire chez *Drosophila melanogaster*, *C. R. Acad. Sci. Paris*, 262, 2106, 1966.
30. **Bussereau, F.**, Étude du symtome de la sensibilité au CO_2 produit par le virus de la stomatite vésiculaire chez "*Drosophila Melanogaster*". II. VSV de sérotype Indiana, *Ann. Inst. Pasteur Paris)*, 122, 1029, 1972.
31. **Bussereau, F.**, Étude du symptome de la sensibilite au CO_2 product par le virus de la stomatite vésiculaire chez *Drosophila melanogaster*. I. VSV de sérotype New Jersey et le virus Cocal, *Ann. Inst. Pasteur Paris*, 121, 223, 1971.

32. **Burnet, F. M. and Galloway, J. A.**, The propagation of the virus of vesicular stomatitis in the chorio-allantoic membrane of the developing hen's egg, *Br. J. Exp. Pathol.*, 15, 105, 1934.
33. **Eichhorn, E. A. and Mathei, C. A.**, Propagation of the virus of vesicular stomatitis in the chick embryo, *J. Am. Vet. Med. Assoc.*, 94, 608, 1939.
34. **Sigurdson, B.**, The influence of age of host and temperature of incubation on infection of the chick embryo with vesicular stomatitis virus, *J. Exp. Med.*, 78, 17, 1943.
35. **Carrel, A., Olitsky, P. K., and Long, P. H.**, Multiplication du virus de la stomatite vesiculaire du cheval dans des cultures de tissues, *C. R. Soc. Biol.*, 98, 827, 1928.
36. **Cox, H. R., Syverton, J. T., and Olitsky, P. K.**, Cultivation of vesicular stomatitis virus, *Proc. Soc. Exp. Biol. Med.*, 30, 896, 1933.
37. **Bachrach, H. L., Callis, J. J., and Hess, W. R.**, The growth and cytopathogenicity of vesicular stomatitis virus in tissue culture, *J. Immunol.*, 75, 186, 1955.
38. **Sellers, R. F.**, Growth and titration of the viruses of foot-and-mouth disease and vesicular stomatitis in kidney monolayer tissue cultures, *Nature (London)*, 176, 547, 1955.
39. **McClain, M. E. and Hackett, A. J.**, A comparative study of the growth of vesicular stomatitis virus in five tissue culture systems, *J. Immunol.*, 80, 356, 1958.
40. **Nishiyama, Y., Ito, Y., Shimokata, K., Kimura, Y., and Nagata, I.**, Polykaryocyte formation induced by VSV in mouse L cells, *J. Gen. Virol.*, 32, 85, 1976.
41. **Takehara, M.**, Polykaryocytosis induced by vesicular stomatitis virus infection in BHK-21 cells, *Arch. Virol.*, 49, 297, 1975.
42. **Chany-Fournier, F., Moreau, M. -C., and Valle Santos, M.**, Induction de polycaryocytes par le virus de la stomatite vêsiculaire dans les cellules de Rat XC, *C. R. Acad. Sci. Paris*, 282, 2007, 1976.
43. **Bachrach, H. L., Callis, J. J., and Hess, W. R.**, Vesicular stomatitis virus in tissue cultures and cell suspensions, *Proc. Soc. Exp. Biol. Med.*, 91, 177, 1956.
44. **Paucker, H., Skurska, Z., and Henle, W.**, Quantitative studies on viral interference in suspended L cells. I. Growth characteristics and interfering activities of vesicular stomatitis, Newcastle disease, and influenza A viruses, *Virology*, 17, 301, 1962.
45. **Padgett, B. L. and Walker, D. L.**, Effect of persistent fibroma virus infection on susceptibility of cells to other viruses, *J. Virol.*, 5, 199, 1970.
46. **Thacore, H. R. and Youngner, J. S.**, Rescue of vesicular stomatitis virus from interferon-induced resistance by superinfection with vaccinia. I. Rescue in cell cultures from different species, *Virology*, 56, 505, 1973.
47. **Edelman, R. and Wheelock, E. F.**, Specific role of each leukocyte type in viral infections. II. Phytohemagglutinin-treated lymphocytes as host cells for vesicular stomatitis virus replication *in vitro*, *J. Virol.*, 2, 440, 1968.
48. **Nowakowski, M., Feldman, J. P., Kano, S., and Bloom, B. R.**, The production of vesicular stomatitis virus by antigen- or mitogen-stimulated lymphocytes and continuous lymphoblastoid lines, *J. Exp. Med.*, 137, 1042, 1973.
49. **Clark, H F., Cohen, M. M., and Karzon, D. T.**, Characterization of reptilian cell lines established at incubation temperatures of 23 to 36°, *Proc. Soc. Exp. Biol. Med.*, 133, 1039, 1970.
50. **Clark, H F. and Karzon, D. T.**, Terrapene heart (TH-1), a continuous cell line from the heart of the box turtle *Terrapene carolina*, *Exp. Cell. Res.*, 48, 263, 1967.
51. **Michalski, F., Cohen, M. M., and Clark, H F.**, Adult and embryonic gecko cells *in vitro*: Growth characteristics, infection by rabies, sindbis and polyoma virusesow, and transformation by SV40, *Proc. Soc. Exp. Biol. Med.*, 146, 337, 1974.
52. **Clark, H F., Cohen, M. M., and Lunger, P. D.**, Comparative characterization of a C-type virus-producing cell line (VSW) and a virus-free cell line (VH2) from *Vipera russelli*, *J. Nat. Cancer Inst.*, 51, 645, 1973.
53. **Shindarov, L. and Sawow, Z.**, Zuchtung des Virus der Vesikularen Stomatitis auf Gewebekultur von Kaltblutern (*Testudo graeca*), *Zentralbl. Bakteriol. Parasitenkd. Infektionskr. Hyg. Abt. Ref.*, 186, 495, 1964.
54. **Solis, J. and Mora, E. C.**, Virus susceptibility range of the fathead minnow (*Pimephales promelas*) poikilothermic cell line, *Appl. Microbiol.*, 19, 1, 1970.
55. **Kelly, R. K. and Loh, P. C.**, Some properties of an established fish cell line from *Xiphophorus helleri* (Red Swordtail), *In Vitro*, 9, 73, 1973.
56. **Lee, M. H. and Loh, P. C.**, Some properties of an established fish cell line from the marine fish, *Caranx mate* (Omaka), *Proc. Soc. Exp. Biol. Med.*, 150, 40, 1975.
57. **Buckley, S. M.**, Susceptibility of the *Aedes albopictus* and *A. aegypti* cell lines to infection with arboviruses, *Proc. Soc. Exp. Biol. Med.*, 131, 625, 1969.
58. **Yang, Y., Stoltz, D. B., and Prever, L.**, Growth of vesicular stomatitis virus in a continuous culture line of *Anthereaea eucalypti* moth cells, *J. Gen. Virol.*, 5, 473, 1969.

59. **Mudd, J. A., Leavitt, R. W., Kingsbury, D. T., and Holland, J. J.,** Natural selection of mutants of vesicular stomatitis virus by cultured cells of *Drosophila melanogaster, J. Gen. Virol.,* 20, 341, 1973.
60. **Franklin, R. M.,** Studies on the growth of vesicular stomatitis virus in tissue culture, *Virology,* 5, 408, 1958.
61. **Cooper, P. P.,** Some characteristics of vesicular stomatitis virus growth-curves in tissue culture, *J. Gen. Microbiol.,* 17, 327, 1957.
62. **Wagner, R. R., Levy, A. H., Snyder, R. M., Ratcliff, G. A., and Hyatt, D. F.,** Biologic properties of two plaque variants of vesicular stomatitis virus (Indiana serotype), *J. Immunol.,* 91, 112, 1963.
63. **Howatson, A. F.,** Vesicular stomatitis and related viruses, *Adv. Virus Res.,* 16, 195, 1970.
64. **Vilcek, J., Yamazaki, S., and Havell, E. A.,** Interferon induction by vesicular stomatitis virus and its role in virus replication, *Infect. Immun.,* 18, 863, 1977.
65. **Henderson, W. M.,** Some observations on the quantitative study of vesicular stomatitis virus, *J. Comp. Pathol. Ther.,* 58, 172, 1948.
66. **Valle, M.,** Factors affecting plaque assay of animal viruses, *Acta Pathol. Microbiol. Scand.,* Suppl. 6, 219, 6, 1971.
67. **Halonen, P. E., Murphy, F. A., Fields, B. N., and Reese, D. R.,** Hemagglutinin of rabies and some other bullet-shaped viruses, *Proc. Soc. Exp. Biol. Med.,* 127, 1037, 1968.
68. **Arstila, P., Halonen, P., and Salmi, A.,** Hemagglutinin of Vesicular Stomatitis Virus, *Arch. Gesamte Virusforsch.,* 27, 198, 1969.
69. **Knudson, D. L.,** Rhabdoviruses, *J. Gen. Virol.,* 20, 105, 1972.
70. **Howatson, A. F.,** Vesicular stomatitis and related viruses, *Adv. Virus. Res.,* 16, 195, 1970.
71. **Bishop, D. H. L., Aaslestad, H. G., Clark, H F., Flamand, A., Obijeski, J. F., Repik, P., and Roy, P.,** Evidence for the sequence homology and genome size of rhabdovirus RNAs, in *Negative Strand Viruses,* Mahy, B. W. J. and Barry, R. D., Eds., Academic Press, New York, 1975, 259.
72. **Cartwright, B. and Brown, F.,** Serological relationships between different strains of vesicular stomatitis virus, *J. Gen. Virol.,* 16, 391, 1972.
73. **Jonkers, A. H., Shope, R. E., Aitken, T. H. G., and Spence, L.,** Cocal virus, a new agent in Trinidad related to vesicular stomatitis virus, type Indiana, *Am. J. Vet. Res.,* 25, 236, 1964.
74. **Bergold, G. H. and Munz, K.,** Characterization of Piry Virus, *Arch. Gesamte Virusforsch.,* 31, 152, 1970.
75. **Thormar, H.,** A comparison of cocal and vesicular stomatitis virus, serotypes New Jersey and Indiana, *Virology,* 31, 323, 1967.
76. **Pringle, C. R. and Wunner, W. H.,** A comparative study of the structure and function of the vesicular stomatitis virus genome, in *Negative Strand Viruses,* Mahy, B. W. J. and Barry, R. D., Eds., Academic Press, New York, 1975, 707.
77. **Artsob, H. and Spence, L.,** Persistent infection of mosquito cell lines with vesicular stomatitis virus, *Acta Virol.,* 18, 331, 1974.
78. **Bussereau, F.,** The CO_2 sensitivity induced by two rhabdoviruses, Piry and Chandipura, in *Drosophila melanogaster, Ann. Microbiol. Paris,* 126 B, 389, 1975.
79. **Galtier, M.,** Deuxieme note sur la sage, *C. R. Acad. Sci. Paris,* 89, 444, 1879.
80. **Fernandes, M. V., Wiktor, T. J., and Koprowski, H.,** Endosymbiotic relationship between animal viruses and host cells. A study of rabies virus in tissue culture, in *Slow, Latent, and Temperate Virus Infections,* Gajdusek, D. C., Gibbs, C. J., Jr., and Alpers, M., Eds., U.S. Public Health Serv. Publ., 1378, 125, 1965.
81. **Van Rooyen, C. E. and Rhodes, A. J.,** *Virus Diseases of Man,* Oxford University Press, London, 1948, 820.
82. **Lepine, P.,** On the evolution of fixed strains of rabies virus, *J. Hyg.,* 38, 180, 1938.
83. **Lepine, P.,** Fermi-type vaccine, in Laboratory Techniques in Rabies, Kaplan, M. M. and Koprowski, H., Eds., World Health Organization, Geneva, 1973, 339.
84. **Selimov, M. A. and Morogova, V. M.,** Phenolized, freeze-dried sheep brain vaccine, A. Method used in the USSR, in Laboratory Techniques in Rabies, Kaplan, M. M. and Koprowski, H., Eds., World Health Organization, Geneva, 1973, chap. 21.
85. **Seligmann, B., Jr.,** Semple-type vaccine, in Laboratory Techniques in Rabies, Kaplan, M. M. and Koprowski, H., Eds., World Health Organization, Geneva, 1973, chap. 19.
86. **Fuenzalida, E.,** Human pre-exposure rabies immunization with suckling mouse brain vaccine, *Bull. W.H.O.,* 46, 561, 1972.
87. **Sikes, R. K. and Larghi, O. P.,** Purified rabies vaccine: Development and comparison of potency and safety with two human rabies vaccines, *J. Immunol.,* 99, 545, 1967.
88. **Karakujumčan, M. K., Pariz, B. M., and Solov'ev, V. D.,** Suckling rat brain vaccine, in Laboratory Techniques in Rabies, Kaplan, M. M. and Koprowski, H., Eds., World Health Organization, Geneva, 1973, chap. 22.

89. **Gispen, R.**, Suckling rabbit brain vaccine, in Laboratory Techniques in Rabies, Kaplan, M. M. and Koprowski, H., Eds., World Health Organization, Geneva, 1973, chap. 24.
90. **Habel, K.**, Factors influencing the efficacy of phenolized rabies vaccines. I. Strains of fixed virus, *Public Health Rep.*, 55, 1619, 1940.
91. **Mason, H. C.**, A comparative study of the behavior of street and fixed viruses of rabies in the mouse, *Public Health Rep.*, 36, 153, 1942.
92. **Murphy, F. A.**, Rabies pathogenesis. Brief review, *Arch. Virol.*, 54, 279, 1977.
93. **Parker, R. L. and Wilsnack, R. E.**, Pathogenesis of skunk rabies virus: quantitation in skunks and foxes, *Am. J. Vet. Res.*, 27, 33, 1966.
94. **Sikes, R. K.**, Pathogenesis of rabies in wildlife. I. Comparative effect of varying doses of rabies virus inoculated into foxes and skunks, *Am. J. Vet. Res.*, 23, 1041, 1962.
95. **Dierks, R. E., Murphy, F. A., and Harrison, A. K.**, Extraneural rabies virus infection. Virus development in fox salivary gland, *Am. J. Pathol.*, 54, 251, 1969.
96. **Constantine, D. G.**, Bat rabies in the southwestern United States, *Public Health Rep.*, 82, 867, 1967.
97. **Baer, G. M. and Bales, G. L.**, Experimental rabies infection in the Mexican freetail bat, *J. Infect. Dis.*, 117, 82, 1967.
98. **Sulkin, S. E., Allen, R., Sims, R., Krutzsch, P. H., and Chansoo, K.**, Studies on the pathogenesis of rabies in insectivorous bats. II. Influence of environmental temperature, *J. Exp. Med.*, 112, 595, 1960.
99. **Kligler, I. J. and Bernkopf, H.**, Cultivation of rabies virus in the allantois of the developing chick embryo, *Proc. Soc. Exp. Biol. Med.*, 39, 212, 1938.
100. **Koprowski, H. and Cox, H. R.**, Studies on chick embryo adapted rabies virus. I. Culture characteristics and pathogenicity, *J. Immunol.*, 60, 533, 1948.
101. **Yoshino, K., Kondo, A., Kuma, N., and Taniguchi, H.**, Infection of the one-day-old fertile hen's egg with rabies virus. VI. Strain differences in the egg adaptability, *Arch. Gesamte Virusforsch.*, 10, 684, 1960.
102. **Yoshino, K., Suzuki, M., and Kondo, A.**, Infection of the one-day-old fertile hen's egg with rabies virus. VII. Comparison of viral yields obtained by inoculation of eggs of different ages with different strains and production of vaccine, *Arch. Gesamte Virusforsch.*, 10, 698, 1960.
103. **Powell, H. M. and Culbertson, C. G.**, Cultivation of fixed rabies virus in embryonated duck eggs, *Public Health Rep.*, 65, 400, 1950.
104. **Hoskins, J. M.**, Duck-embryo vaccine, in Laboratory Techniques in Rabies, Kaplan, M. M. and Koprowski, H., Eds., World Health Organization, Geneva, 1973, chap. 27.
105. **Noguchi, H.**, Contribution to the cultivation of the parasite of rabies, *J. Exp. Med.*, 18, 314, 1913.
106. **Levaditi, C.**, Virus rabique et culture des cellules "in vitro", *C. R. Soc. Biol.*, 75, 505, 1913.
107. **Webster, L. T. and Chow, A. D.**, Propagation of rabies virus in tissue culture and the successful use of culture virus as an antirabic vaccine, *Science*, 84, 487, 1936.
108. **Bernkopf, H. and Kligler, I. J.**, The cultivation of rabies virus in tissue culture, *Br. J. Exp. Pathol.*, 18, 481, 1937.
109. **Schultz, E. W. and Williams, G. F.**, Cultivation of rabies virus, *Proc. Soc. Exp. Biol. Med.*, 37, 372, 1937.
110. **Atanasiu, P. and Lauent, M. C.**, Multiplication du virus de la rage des mus sur une cellule gliale (Ependymome de la Souris) in culture de tissues, *C. R. Acad. Sci.*, 245, 2562, 1957.
111. **Vieuchange, J., Vialat, C., Gruest, J., and Bequignon, R.**, Essais de culture *in vitro* du virus rabique des rues, *Ann. Inst. Pasteur Paris*, 90, 361, 1956.
112. **Kissling, R. E.**, Growth of rabies virus in non-nervous tissue culture, *Proc. Soc. Exp. Biol. Med.*, 98, 223, 1958.
113. **Kondo, A.**, Growth characteristics of rabies virus in primary chick embryo cell, *Virology*, 27, 199, 1965.
114. **Sokol, F., Kuwert, E., Wiktor, T. J., Hummeler, K., and Koprowski, H.**, Purification of rabies virus grown in tissue culture, *J. Virol.*, 2, 836, 1968.
115. **Iwasaki, Y. and Clark, H F.**, Rabies virus infection in mouse neuroblastoma cells, *Lab. Invest.*, 36, 578, 1977.
116. **Clark, H F.**, Rabies viruses increase in virulence when propagated in neuroblastoma cell culture, *Science*, 199, 1072, 1978.
117. **Lang, R., Petermann, H. G., Branche, R., and Soulebot, J. P.**, Adaptation d'un virus rabique fixe aux cellules vénales de singes. Utilisation de ce virus pour la préparation d'un vaccin antirabique inactivé, *C. R. Acad. Sci. Paris*, 269, 2287, 1969.
118. **Hronovsky, V., Benda, R., and Cinatl, J.**, Propagation of street rabies virus in primary cultures of dog kidney cells, *Acta Virol.*, 10, 181, 1966.
119. **Abelseth, M. K.**, Propagation of rabies virus in pig kidney cell culture, *Can. Vet. J.*, 5, 84, 1964.

120. **Wiktor, T. J., Fernandes, M. V., and Koprowski, H.,** Cultivation of rabies virus in human diploid cell strain WI-38, *J. Immunol.*, 93, 353, 1964.
121. **Clark, H F.,** Growth and attenuation of rabies virus in cell cultures of reptilian origin, *Proc. Soc. Exp. Biol. Med.*, 139, 1317, 1972.
122. **Atanasiu, P., Raynaud, J., and Raynaud, A.,** Développement du virus rabique sur des cellules embryonnaires de Reptiles, *C. R. Acad. Sci. Paris*, 276, 2097, 1973.
123. **Yoshino, K., Taniguchi, S., and Arai, K.,** Autointerference of rabies virus in chick embryo fibroblasts, *Proc. Soc. Exp. Biol. Med.*, 123, 387, 1966.
124. **Iwasaki, Y., Wiktor, T. J., and Koprowski, H.,** Early events of rabies virus replication in tissue cultures. An electron microscopic study, *Lab. Invest.*, 28, 142, 1973.
125. **Kuwert, E., Wiktor, T. J., Sokol, F., and Koprowski, H.,** Hemagglutination by rabies virus, *J. Virol.*, 2, 1381, 1968.
126. **Kaplan, M. M., Wiktor, T. J., Maes, R. F., Campbell, J. B., and Koprowski, H.,** Effect of polyions on the infectivity of rabies virus in tissue culture: Construction of a single-cycle growth curve, *J. Virol.*, 1, 145, 1967.
127. **Clark, H F. and Wiktor, T. J.,** Temperature-sensitivity characteristics distinguishing substrains of fixed rabies virus: Lack of correlation with plaque-size markers or virulence for mice, *J. Infect. Dis.*, 125, 637, 1972.
128. **Casals, J.,** Influence of age factors on susceptibility of mice to rabies virus, *J. Exp. Med.*, 72, 445, 1940.
129. **Koprowski, H.,** The mouse inoculation test, in Laboratory Techniques in Rabies, Kaplan, M. M. and Koprowski, H., Eds., World Health Organization, Geneva, 1973, chap. 7.
130. **Webster, L. T. and Dawson, J. R., Jr.,** Early diagnosis of rabies by mouse inoculation. Measurement of humoral immunity to rabies by mouse protection test, *Proc. Soc. Exp. Biol. Med.*, 32, 570, 1935.
131. **Bijlenga, G. and van den Bogaard, A. E. J. M.,** In vivo enhancement of rabies infection by diethylaminoethyl-dextran, *Arch. Gesamte Virusforsch.*, 42, 96, 1973.
132. **Enright, J. B., Goggin, J. E., Frye, F. L., Franti, C. E., and Behymer, D. E.,** Effects of corticosteroids on rabies virus infections in various animal species, *J. Am. Vet. Med. Assoc.*, 156, 765, 1970.
133. **Bell, J. F. and Moore, G. J.,** Effects of high ambient temperature on various stages of rabies virus infection in mice, *Infect. Immun.*, 10, 510, 1974.
134. **Bell, J. F., Clark, H F., and Moore, G. J.,** Differences in efficiency of protective effect caused by high ambient temperature in mice infected with diverse substrains of rabies virus, *J. Gen. Virol.*, 36, 307, 1977.
135. **Larghi, O. P., Nebel, A. E., Lazaro, L., and Savy, V. L.,** Sensitivity of BHK/21 cells supplemented with diethylaminoethyl-dextran for detection of street rabies virus in saliva samples, *J. Clin. Microbiol.*, 1, 243, 1975.
136. **Smith, A. L., Tignor, G. H., Mifune, K., and Motohashi, T.,** Isolation and assay of rabies serogroup viruses in CER cells, *Intervirology*, 8, 92, 1977.
137. **Smith, A. L., Tignor, G. H., Emmons, R. W., and Woodie, J. D.,** Isolation of field rabies virus strains in CER and murine neuroblastoma cell cultures, *Intervirology*, 9, 359, 1978.
138. **Debbie, J. G., Andrulonis, J. A., and Abelseth, M. K.,** Rabies antibody determination by immunofluorescence in tissue culture, *Infect. Immun.*, 5, 902, 1972.
139. **Smith, J. D., Yager, P. A., and Baer, G. M.,** A rapid reproducible test for determining rabies neutralizing antibody, *Bull. W.H.O.*, 48, 535, 1973.
140. **Sedwick, W. D. and Wiktor, T. J.,** Reproducible plaquing system for rabies, lymphocytic choriomeningitis, and other ribonucleic acid viruses in BHK-21/13S agarose suspensions, *J. Virol.*, 1, 1224, 1967.
141. **Wiktor, T. J. and Clark, H F.,** Application of the plaque assay technique to the study of rabies virus-neutralizing antibody interactions, *Ann. Microbiol. (Paris)*, 124 A, 271, 1973.
142. **Strating, A., Mitzel, J. R., and Slater, E. A.,** Plaque assay of rabies virus on porcine kidney cell monolayers, *J. Clin. Microbiol.*, 2, 126, 1975.
143. **Schneider, L.,** Cell monolayer plaque test, in Laboratory Techniques in Rabies, Kaplan, M. M. and Koprowski, H., Eds., World Health Organization, Geneva, 1973, 339.
144. **Yoshino, K., Taniguchi, S., and Asai, K.,** Plaque assay of rabies virus in chick embryo cells, *Arch. Gesamte Virusforsch.*, 18, 370, 1966.
145. **Atanasiu, P.,** Animal inoculation and the Negri body, in *The Natural History of Rabies*, Vol. 1, Baer, G. M., Ed., Academic Press, New York, 1975, chap. 19.
146. **Kissling, R. E.,** The fluorescent antibody test in rabies, in *The Natural History of Rabies*, Vol. 1, Baer, G. M., Ed., Academic Press, New York, 1975, chap. 20.
147. **Murphy, F. A., Halonen, P. E., Gary, G. M., and Reese, D. R.,** Physical characterization of rabies virus hemagglutinin, *J. Gen. Virol.*, 3, 289, 1968.

148. **Halonen, P. E., Toivanen, P., and Nikkari, T.,** Non-specific inhibitors of activity of hemagglutinins of rabies and vesicular stomatitis viruses, *J. Gen. Virol.*, 22, 309, 1974.
149. **Aaslestad, H. G.,** Antigenic composition of rabies virus, in *The Natural History of Rabies,* Vol. 1, Baer, G. M., Ed., Academic Press, New York, 1975, chap. 3.
150. **Wiktor, T. J., Kuwert, E., and Koprowski, H.,** Immune lysis of rabies virus-infected cells, *J. Immunol.*, 101, 1271, 1968.
151. **Iwasaki, Y., Gerhard, W., and Clark, H F.,** Role of host immune response in the development of either encephalitis or paralytic disease after experimental rabies infection in mice, *Infect. Immun.*, 18, 220, 1977.
152. **Shope, R.,** Rabies virus antigenic relationships, in *The Natural History of Rabies,* Vol. 1, Baer, G. M., Ed., Academic Press, New York, 1975, chap. 8.
153. **Clark, H F. and Wiktor, T. J.,** Plasticity of phenotypic characters of rabies-related viruses: spontaneous variation in the plaque morphology, virulence, and temperature-sensitivity characters of serially propagated Lagos bat and Mokola viruses, *J. Infect. Dis.*, 130, 608, 1974.
154. **Buckly, S. J.** Singh's *Aedes albapictus* cell cultures as helper cells for the adaptation of Obodhiang and Kotankan viruses of the rabies subgroup to some vertebrate cell cultures, *Appl. Microbiol.*, 25, 695, 1973.
155. **Van der Westhuizen, B.,** Studies on bovine ephemeral fever. I. Isolation and preliminary characterization of a virus from natural and experimentally produced cases of bovine ephemeral fever, *Onderstepoort J. Vet. Res.*, 34, 29, 1967.
156. **Sasaki, N., Kodama, K., Iwamoto, I., Izumida, A., and Matsubara, T.,** Serial transmission in suckling mice of a virus from cattle with bovine epizootic fever, *Jpn. J. Microbiol.*, 12, 251, 1968.
157. **Doherty, R. L., Standfast, M. A., and Clark, I. A.,** Adaptation to mice of the causative virus of ephemeral fever of cattle from an epizootic in Queensland, *Austr. J. Sci.*, 31, 365, 1969.
158. **Spradbow, P. B. and Francis, J.,** Observations on bovine ephemeral fever and isolation, *Austr. Vet. J.*, 45, 525, 1969.
159. **Inaba, Y., Tanaka, Y., Sato, K., Ito, H., Osori, T., and Matumoto, M.,** Propagation in laboratory animals and cell cultures of a virus from cattle with bovine epizootic fever, *Jpn. J. Microbiol.*, 12, 253, 1968.
160. **Tzipori, S. and Spradbow, P. B.,** Development and behaviour of a strain of bovine ephemeral fever virus with unusual host range, *J. Comp. Pathol.*, 84, 1, 1974.
161. **Matumoto, M., Inaba, Y., Tanaka, Y., Ito, H., and Omori, T.,** Behavior of bovine ephemeral fever virus in laboratory animals and cell cultures, *Jpn. J. Microbiol.*, 14, 413, 1970.
162. **Heuschele, W. P.,** Bovine ephemeral fever. I. Characteristics of the causative virus, *Arch. Gesamte Virusforsch.*, 30, 195, 1970.
163. **Snowdon, W. A.,** Bovine ephemeral fever: The reaction of cattle to different strains of ephemeral fever virus and the antigenic comparison of two strains of virus, *Austr. Vet. J.*, 46, 258, 1970.
164. **Tzipori, S.,** The isolation of bovine ephemeral fever virus in cell cultures and evidence for autointerference, *Austr. J. Exp. Biol. Med. Sci.*, 53, 273, 1975.
165. **Tzipori, S.,** Plaque assay and characteristics of strains of bovine ephemeral fever virus in vero cells, *Br. Vet. J.*, 131, 720, 1975.
166. **Murphy, F. A. and Fields, B. N.,** Kern Canyon Virus: Electron microscopic and immunological studies, *Virology*, 33, 625, 1967.
167. **Aaslestad, H. G., Clark, H F., Bishop, D. H. L., and Koprowski, H.,** Comparison of the ribonucleic acid polymerases of two rhabdoviruses, Kern Canyon virus and Vesicular Stomatitis virus, *J. Virol.*, 7, 726, 1971.
168. **Karabatsos, N., Lipman, M. B., Garrison, M. S., and Mongillo, C. A.,** The morphology, morphogenesis, and serological characterization of the rhabdoviruses Navarro, Kwatta, and Mossuril, *J. Gen. Virol.*, 21, 429, 1973.
169. **Murphy, F. A., Johnson, H. N., Harrison, A. K., and Shope, R. E.,** Ultrastructural characterization of a new member of the rhabdovirus group-Klamath virus, *Arch. Gesamte Virusforsch.*, 37, 323, 1972.
170. **Whitney, E.,** Flanders strain, an arbovirus newly isolated from mosquitos and birds of New York State, *Am. J. Trop. Med. Hyg.*, 13, 123, 1964.
171. **Murphy, F. A., Coleman, P. H., and Whitfield, S. D.,** Electron microscopic observations of Flanders virus, *Virology*, 30, 314, 1966.
172. **Jenson, A. B., Rabin, E. R., Wende, R. D., and Melnick J. L.,** A comparative light and electron microscopic study of rabies and Hart Park virus encephalitis, *Exp. Mol. Pathol.*, 7, 1, 1967.
173. **Schaperclaus, W.,** *Fischkrankheiten,* 3rd ed., Akademie Verlag, Berlin, 1954, 537.
174. **Rasmussen, C. J.,** A biological study of the Egtved disease (INUL), *Ann. N. Y. Acad. Sci.*, 126, 427, 1965.

175. **Ghittino, P.**, Viral hemorrhagic septicemia (VHS) in rainbow trout in Italy, *Ann. N. Y. Acad. Sci.*, 126, 468, 1965.
176. **Jensen, M. H.**, Preparation of fish tissue cultures for virus research, *Bull. Off. Int.*, 59, 131, 1963.
177. **Pfitzner, I.**, Beitrag zur Atiologie der "Haemorrhagischen Virus-septikaemie der Regenbogenforellen," *Zentralbl. Bakteriol. Parasitenk. Infektionskr.*, 201, 306, 1966.
178. **Jensen, M. H.**, Research on the virus of Egtved disease, *Ann. N. Y. Acad. Sci.*, 126, 422, 1965.
179. **de Kinkelin, P. and Scherrer, R.**, Le virus d'Egtved. I. Stabilité, dêveloppment et structure du virus de la souche danoise F_1, *Ann. Rech. Vet.*, 1, 17, 1970.
180. **Campbell, J. B. and Wolf, K.**, Plaque assay and some characteristics of Egtved virus (virus of viral hemorrhagic septicemia of rainbow trout), *Can. J. Microbiol.*, 15, 635, 1969.
181. **Vestergård Jørgensen, P. E. and Meyling, A.**, Egtved virus: Demonstration of virus antigen by the fluorescent antibody technique in tissues of rainbow trout affected by viral haemorrhagic septicaemia and in cell cultures infected with Egtved virus, *Arch. Gesamte Virusforsch.*, 36, 115, 1972.
182. **Darlington, R. W., Trafford, R., and Wolf, K.**, Fish rhabdoviruses: Morphology and ultrastructure of North American Salmonid isolates, *Arch. Gesamte Virusforsch.*, 39, 257, 1972.
183. **Parisot, T. J., Yasutake, W. T., and Klontz, G. W.**, Virus diseases of the Salmonidae in western United States. I. Etiology and epizootiology, *Ann. N. Y. Acad. Sci.*, 126, 502, 1965.
184. **Klontz, G. W., Yasutake, W. T., and Parisot, T. J.**, Virus diseases of the Salmonidae in western United States. III. Immunopathological aspects, *Ann. N. Y. Acad. Sci.*, 126, 531, 1965.
185. **Amend, D. F.**, Control of infectious hematopoietic necrosis virus disease by elevating the water temperature, *J. Fish. Res. Board Can.*, 27, 265, 1970.
186. **Wingfield, W. H., Fryer, J. L., and Pilcher, K. S.**, Properties of the Sockeye Salmon virus (Oregon strain), in *Proc. Soc. Exp. Biol. Med.*, 130, 1055, 1969.
187. **Nims, L., Fryer, J. L., and Pilcher, K. S.**, Studies of replication of four selected virusesn two cell lines derived from Salmonid fish, *Proc. Soc. Exp. Biol. Med.*, 135, 6, 1970.
188. **Amend, D. F. and Chambers, V. C.**, Morphology of certain viruses of Salmonid fishes. I. *In vitro* studies of some viruses causing hematopoietic necrosis, *J. Fish. Res. Board Can.*, 27, 1285, 1970.
189. **McAllister, P. E. and Pilcher, K. S.**, Autointerference in infectious hematopoietic necrosis virus of Salmonid fish, *Proc. Soc. Exp. Biol. Med.*, 145, 840, 1974.
190. **McAllister, P. E., Fryer, J. L. and Pilcher, K. S.**, Further characterization of infectious hematopoietic necrosis virus of Salmonid fish (Oregon strain), *Arch. Gesamte Virusforsch.*, 44, 270, 1974.
191. **McCain, B. B., Fryer, J. L., and Pilcher, K. S.**, Antigenic relationships in a group of three viruses of Salmonid fish by cross neutralization, *Proc. Soc. Exp. Biol. Med.*, 137, 1042, 1971.
192. **Fijan, N. N.**, Infectious dropsy in carp — a disease complex, *Symp. Zool. Soc. London*, 30, 39, 1972.
193. **Fijan, N. N., Petrinec, Z., Sulimanović, D., and Zwillenberg, L. O.**, Isolation of the viral causative agent from the acute form of infectious Dropsy of carp, *Vet. Arh.*, 41, 125, 1971.
194. **de Kinkelin, P. and Le Berre, M.**, Rhabdovirus des poissons. II. Proprietes *in vitro* du virus de la virêmie printanière de la carpe, *Ann. Microbiol. Paris*, 125 A, 113, 1974.
195. **Bachmann, P. A. and Ahne, W.**, Biological properties and identification of the agent causing swim bladder inflammation in carp, *Arch. Gesamte Virusforsch.*, 44, 261, 1974.
196. **de Kinkelin, P., Galimard, B., and Bootsma, R.**, Isolation and identification of the causative agent of "Red Disease" of pike, *Nature (London)*, 241, 465, 1973.
197. **de Kinkelin, P., Le Berre, M., and Lenoir, G.**, Rhabdovirus des poissons. I. Propriêtés *in vitro* du virus de la maladie rouge de l'alevin de brochet, *Ann. Microbiol. Paris*, 125 A, 93, 1974.
198. **Bussereau, F., de Kinkelin, P., and Le Berre, M.**, Infectivity of fish rhabdoviruses for drosoplila melanogaster, *Ann. Microbiol. Paris*, 126 A, 389, 1975.
199. **Wolf, K. and Quimby, M. C.**, Fish viruses: buffers and methods for plaquing eight agents under normal atmosphere, *Appl. Microbiol.*, 25, 659, 1973.
200. **Hill, B. J.**, Physico-chemical and serological characterization of five rhabdoviruses infecting fish, *J. Gen. Virol.*, 27, 369, 1975.
201. **Clark, H F.**, unpublished data.
202. **Wiktor, T. J.**, unpublished results.
203. **Clark, H F.**, unpublished results.

Chapter 3

SEROLOGIC RELATIONSHIPS OF ANIMAL RHABDOVIRUSES

Christina L. Frazier and Robert E. Shope

TABLE OF CONTENTS

I. INTRODUCTION

A. Viruses

The rhabdoviruses of vertebrates and invertebrates are divided in this chapter into three categories based on host range. The majority of the viruses are propagated in the laboratory mouse and are thus designated rhabdoviruses of warm-blooded vertebrates despite the fact that many also have arthropod hosts, and that some are found in nature so far only in arthropods. The second category contains six viruses that are isolated from fish and grown in fish or fish cell lines. The final category is made up of Sigma virus and viruses from entamoeba and the blue crab. These replicate in invertebrate animals only, have not been shown to infect vertebrate animals, and for lack of substantial serological comparisons to other rhabdoviruses, will not be considered further here.

B. Serological Techniques

The procedures which have been applied in studies of the serological relationships of the rhabdoviruses comprise: complement fixation (CF); hemagglutination inhibition (HI); immunofluorescence (FA), both direct and indirect; agar gel diffusion; vaccination challenge; and radioimmune assay (RIA); in addition to a battery of neutralization tests including plaque reduction (PRN) in tissue culture; fluorescent focus inhibition (FFIT); and serum dilution-constant virus, and virus dilution-constant serum with results expressed as the log neutralization index (LNI). Detailed description of these tests are found in References 1 to 3 and in the literature cited when specific viruses are discussed.

Despite the availability of a large number of viruses and numerous test systems, most of the studies have been concentrated on a few viruses with apparent human and veterinary medical importance and have employed a narrow range of tests often chosen because they are economical, reliable, and simple. Thus the vesicular stomatitis and rabies serogroups have been extensively studied with a broad spectrum of procedures, while many of the other rhabdoviruses of warm-blooded vertebrates have been compared serologically by CF only. The rhabdoviruses of fish with rare exceptions have not been compared virologically with rhabdoviruses of warm-blooded animals although plaque-reduction neutralization tests offer a feasible means of studying both categories.

C. Structure-Function Associations

Studies of Dietzschold, Schneider, and Cox[4] of the three major proteins of vesicular stomatitis, type Indiana virus indicated that not all proteins were detected by all serological tests. The envelope G protein had a molecular weight (MW) of 67,000 daltons and it induced immunity. Anti-G was positive for neutralizing, immunofluorescence, CF, and precipitating activity. The core N protein had a MW of 52,000 daltons and was responsible for group reactivity. Anti-N was devoid of neutralizing activity, but was positive for immunofluorescence, CF, and precipitating activity. The envelope M protein had a MW of 25,000 daltons. Anti-M was negative in the neutralization and immunofluorescence tests, but positive in CF and precipitation tests. Cartwright and Brown[5] also presented evidence that the vesicular stomatitis virus (VSV) G-protein was type specific, while the N protein was more group reactive; and Schneider et al.[6] showed the same relationships in the rabies serogroup. The serological reactions and specificities of L and NS proteins are not known.

The extent of observed antigenic relationship between two viruses depends on several factors in addition to the chemical similarities and differences of the antigenic sites of the viruses. The particular test, the technique for carrying out the test, the potency of the antigen and antibody, the relative proportions of IgM and IgG in sera, and the ratio of G, M, and N in the antigen preparation, or anti-G, anti-N, or anti-M in the antibody preparation can all affect experimental outcome, and on some occasions may lead to inconsistencies in data.

II. SEROLOGICAL RELATIONSHIPS OF RHABDOVIRUSES OF WARM-BLOODED VERTEBRATES

A. Vesicular Stomatitis Serogroup

The *Vesiculovirus* genus corresponds to the VSV serogroup which is divided into three complexes. VS, type Indiana, Cocal, and Alagoas viruses constitute one serocomplex, VSV, type New Jersey (alone) a second, and Piry, Isfahan, and Chandipura viruses constitute the third complex. The division into three complexes is supported by neutralization, CF, and fluorescent focus inhibition test results.

The method of neutralization testing strongly influences the degree of cross-reaction. Shope[7] used undiluted hyperimmune sera with varying dilutions of virus and demonstrated marked cross-reactions between VS, type Indiana and Cocal viruses, and between Chandipura and Piry viruses (Table 1).* Cartwright and Brown[8] using the same test but with sera diluted 1:100 showed less cross-reaction between VS, type Indiana and Cocal viruses, and none at all between Piry and Chandipura viruses. The neutralization test cross-reactions among VS, types Indiana and New Jersey, Cocal and Alagoas (also called "Brazil") were enhanced by using skeletons of tween-ether disrupted virions, indicating that the group reactivity resides primarily in the nucleocapsid. Federer et al.[9] first related Alagoas (Brazil) virus to VS, type Indiana and Cocal viruses. They showed by serum-dilution and by virus-dilution neutralization tests that Alagoas virus differed about equally from VS, type Indiana and from Cocal. They also demonstrated that the cross-reactions became undetectable at serum dilutions beyond 1:40.

The plaque-reduction neutralization test is very sensitive and quite specific; in spite of this specificity, the cross-reactions among VS, type Indiana, Cocal, and Alagoas, as well as that between Chandipura and Piry were shown by Tesh et al. (Table 2).[10]

CF test results obtained in the authors' laboratory (Table 3) and by Federer et al.[9] (Table 4) were consistent with those of others,[8,10,11] and corresponded in specificity very closely with neutralization test results. The CF tests confirm the existence of three serocomplexes in the *Vesiculovirus* genus.

Results of qualitative fluorescent-focus inhibition tests done by the senior author also correlated well with neutralization test results, but showed a broader cross-reaction among the complexes which linked six of the vesiculoviruses (Alagoas was not included) into a serogroup (Table 5).

Cross-challenge tests[9] in cattle used vesicular lesions as the indicator of disease. VS, type Indiana, Alagoas, and Cocal viruses were distinct although some protection was afforded by Alagoas infection to VS, type Indiana challenge, and considerable protection was noted in animals infected with an Argentine strain of Cocal virus and challenged with VS, type Indiana virus. Two of four animals infected with prototype Cocal virus were protected at a low level against Alagoas challenge. The outcome of challenge in general correlated with levels of neutralizing antibody (unlike cross-protection test results with lyssaviruses [see next section]).

The hemagglutination-inhibition[12] and the immunodiffusion[11,13] tests have also been

* All tables appear following text.

successfully applied to the study of the vesiculoviruses, and the results correlated well with those of other serological techniques.

Reichman et al.[14] proposed that a classification system based primarily on RNA oligonucleotide mapping and base sequence homology be weighed heavily with the neutralization test results in differentiating subtypes of VSV New Jersey. They interpreted their data as showing that there are two subtypes of VSV New Jersey, although on the basis of their neutralization test results alone, their subtypes might more appropriately be classified as geographic variants.

B. Rabies Serogroup

The rabies serogroup contains three serocomplexes.[7] The members of the genus *Lyssavirus* (rabies, Mokola, Duvenhage, and Lagos bat viruses) comprise one serocomplex while Obodhiang and kotonkan viruses are lone serotypes in the other two complexes. Buckley[15] has suggested that Mokola virus acts as a serological and biological bridge between the arthropod-associated (Obodhiang and kotonkan) serotypes and the non-arthropod-associated (rabies, Duvenhage, and Lagos bat) serotypes. This hypothesis is supported by a number of the serological test results in this section.

Neutralization tests using undiluted antibody to neutralize varying dilutions of virus in baby mice (Tables 6 and 7) showed that rabies virus was strongly neutralized by Mokola, Lagos bat, and Duvenhage antibody and minimally by kotonkan and Obodhiang antibody.[7,16] In serum-dilution neutralization tests in mice minimal cross-reactions were noted (Tables 8 and 9).[16,17] The Duvenhage immune ascitic fluid-rabies antigen combination exhibited the highest level of cross-reactivity. As with the vesiculoviruses, rabies serogroup, cross-reactions were best demonstrated by using concentrated antibody.

Plaque-reduction neutralization test results confirmed the relationship of Duvenhage, Mokola, and rabies viruses, and demonstrated a cross-reaction between kotonkan and Obodhiang (Table 10).[15] Mokola virus reacted with both arthropod-associated and nonassociated viruses in this test system.

Vaccination-challenge experiments were performed in order to study the relationship among the lyssaviruses and yielded data of potential public health importance. Although rabies virus was neutralized by Lagos bat, Mokola, and Duvenhage hyperimmune mouse ascitic fluids in vitro, vaccination with these viruses did not protect quantitatively against rabies virus challenge (Tables 11 and 12).[16,18] Thus animals may have antibody capable of neutralizing a heterologous virus in vitro and yet not be protected from challenge with this virus.

By CF test, the lyssaviruses (rabies, Mokola, Lagos bat, and Duvenhage) were relatively closely related, and Obodhiang and kotonkan were only distantly related to other members of the group; in tests done in the authors' laboratory the kotonkan reaction with Mokola virus was not always detected (Tables 13 and 14), an inconsistency which is not related to the homologous potency of the Mokola antibody. Cross-reactions between rabies virus itself and Obodhiang or kotonkan were not observed.[19] Schneider et al.[6] have proposed that the group-specific antigen responsible for the cross-reactions of lyssaviruses is the ribonucleoprotein.

The direct FA technique demonstrated the same cross-reactions among rabies serogroup viruses as the CF test (Tables 15 and 16).[16,20] The cross-reactions among rabies, Mokola, Lagos bat, and Duvenhage viruses were consistently observed, however these could be missed at or near terminal dilutions of conjugated antibody. This was apparently the case in a diagnostic setting in South Africa, and led to the recognition of Duvenhage virus as an unusual rabies-like virus.[16] The indirect FA test was more sensitive than the direct, and demonstrated the definite relationship of Obodhiang and kotonkan viruses to Mokola virus (Table 17). Bauer and Murphy[20] showed that this cross-reaction was not present by surface immunofluorescence, indicating that the cross-reacting antigen was probably not viral surface glycoprotein.

By hemagglutination-inhibition tests, Halonen et al.[12] were not able to show cross-reactions between rabies and vesiculoviruses; slight inhibition of Kern Canyon hemagglutinin by rabies antiserum was deemed anomalous and was credited possibly to nonspecific inhibitors (Table 18). This explanation seems likely since extensive experience in our own laboratory indicated that nonspecific inhibitors of rhabdovirus hemagglutinins were not uniformly removed by kaolin, and that the antigens were very sensitive to small residua of these inhibitors.

Radioimmunoassay (RIA) was used to examine rabies, Mokola and Lagos bat viruses for antigenic relationships. Competition experiments among the radiolabeled ribonucleoproteins and the three unlabeled ribonucleoproteins were analyzed via regression lines, with a steep slope indicating a high level of relatedness. Some antigenic overlap was noted among the three viruses but each was distinct (Table 19).[21]

C. Minor Serogroups

The remaining serogroups have been studied in our laboratory mainly by the CF test and are composed of two or three serotypes each (Table 20). Flanders and Hart Park viruses were very closely related by neutralization test (Table 1) and can be distinguished by Ouchterlony gel double diffusion tests.[22] The recent addition of an unpublished virus from the Brazilian Amazon, Mosqueiro, to this serogroup is based on CF results only. The CF results are consistent with considering Hart Park and Flanders as two serotypes within a serocomplex, while Mosquiero is a lone serotype in a second serocomplex.

The cross-reaction observed between Kamese and Mossuril by the neutralization test was much less than that observed by CF.[23] The Kamese serum had a homologous log neutralizing index (LNI) of > 3.5 with heterologous LNI of 1.7. The Mossuril serum homologous LNI was > 3.0 and heterologous, 0.5.

The minor CF cross-reaction between Sawgrass and New Minto viruses was confirmed by plaque-reduction neutralization test.[23] The Sawgrass serum titered 1:320 with itself and 1:10 with New Minto virus. The New Minto serum had a homologous titer of 1:160 and heterologous, 1:40.

The close relationship of Kwatta and BeAn 157575, and the relationship of Timbo and Chaco[25] have been determined by CF test only.

D. Ungrouped Serotypes

Fourteen serotypes (Table 21) have been shown in our laboratory to be unrelated to all other rhabdoviruses of vertebrates by CF testing. As more potent reagents or more sensitive tests are applied, it is of course possible that some of these agents will be grouped. Bovine ephemeral fever virus was not tested because of the restriction on its use in U.S. laboratories, but no serological relationship with any other rhabdovirus has been reported.[6,26] Parry Creek virus, an unpublished rhabdovirus from Australian mosquitoes isolated by Stanley was not included in these studies because a satisfactory homologous CF system was not available.

III. SEROLOGICAL RELATIONSHIPS OF RHABDOVIRUSES OF FISH

A. Spring Viremia of Carp Serogroup

Spring viremia of carp (SVC) virus is indistinguishable from swim-bladder inflammation (SBI) virus and both are distantly related to infectious haematopoetic necrosis

(IHN) virus and to pike fry disease (PFD) virus (Table 22).[27] The virus of infectious haematopoetic necrosis is identical serologically to Oregon sockeye salmon virus and almost identical to Sacramento River Chinook virus.[28]

B. Other Rhabdoviruses of Fish

The three viruses in the spring viremia of carp serogroup differ by neutralization test from Egtved virus, the causal agent of viral hemorrhagic septicemia. Jorgensen[29] showed that 72 of 76 Egtved strains were neutralized by anti-Egtved serum; however four strains were poorly neutralized. These may represent either strains identical to Egtved but difficult to neutralize, or a second serotype of Egtved virus. The distinction could be made by reciprocal tests, but these have not yet been done.

Other rhabdoviruses of fish have been incompletely studied; infectious dropsy of carp virus differs from Egtved,[29] and a grass carp isolate differs serologically from spring viremia of carp virus.[30] It is premature however to consider these as unique serotypes until they are tested with other fish rhabdoviruses. The same holds true for infectious pancreatic necrosis virus which was shown to differ serologically from Egtved, infectious haematopoetic necrosis, and infectious dropsy of carp viruses.[29]

Antisera for fish rhabdoviruses are readily prepared; however except for spring viremia of carp virus, the nucleocapsid antigen of which is distinct by CF test from similar antigens of vesicular stomatitis, rabies, Lagos bat and Mokola viruses,[31] the rhabdoviruses of fish have not been compared serologically with rhabdoviruses of other vertebrate animals.

IV. THE SIGNIFICANCE OF SEROLOGICAL CLASSIFICATION

Means of classification of rhabdoviruses other than serological have been proposed and will undoubtedly be used more frequently in the future as morphologic and physico-chemical methods become available in laboratories where new rhabdoviruses are being isolated. Morphologic groupings are discussed in Chapter 4 in this volume. For the most part, members of a serogroup are alike morphologically and morphogenetically. In the rabies serogroup, however, Obodhiang and kotonkan differ prominently from other members of the group;[20] some of their particles are cone shaped and resemble bovine ephemeral fever and Marco viruses[25] more closely than they resemble rabies virus. This dichotomy has puzzled the taxonomists.

Groupings by RNA structure or protein composition are discussed in Chapters 5 to 9. Reichman et al.[14] have proposed that nucleotide base sequence homology, oligonucleotide maps of virion RNA, and interference by defective-interfering particles be used as criteria in addition to serologic tests in the classification of the New Jersey serotype of vesicular stomatitis virus. These and other nonserologic classification systems should be used where they increase our understanding of epidemiology, pathogenicity, or evolution of rhabdoviruses. Viruses so classified, however, would no longer be "serotypes".

The serological classification of the rabies serogroup has immediate public health implications for parts of Africa and potential implications for other areas where rabies is present. In Africa the standard FA test using a single antibody dilution is no longer sufficient to establish that an isolate is rabies, since Mokola and Lagos bat viruses also react positively. The presence of antigenically related viruses in an area leads to several concerns when vaccination programs are being undertaken and may be the cause of some vaccine failures. Is the proper virus being vaccinated against? Will immunity to or current infection with a related virus result in disease despite vaccination against rabies? At this time these considerations are only potential and are limited to Africa.

Continued serological evaluation of new isolates is essential to detect new rabies-related viruses should they exist and to determine the full geographic and host ramifications of the rabies serogroup viruses. The recognition of vesicular stomatitis viruses in areas where foot-and-mouth disease occurs, or might be introduced, is best done serologically and continues to be of high priority to prevent unnecessary vaccination or slaughter campaigns.

The investigation of the serological relationships among the rhabdoviruses isolated from fish has shown that viruses thought to be distinct entities were serologically indistinguishable. Spring viremia of carp and swim bladder inflammation were treated as separate unrelated viruses because of differences in seasonal occurrence, geographical distribution, and pathogenesis; but serological testing has shown them to be the same or very closely related viruses. The Oregon strain of sockeye salmon virus and infectious haematopoietic necrosis virus, which was isolated from rainbow trout, were at one point believed to be very different viruses, but serological testing showed them to be antigenically indistinguishable.

The serological classification has also established a basis for hypotheses of evolution. Other techniques such as RNA oligonucleotide mapping or RNA homology may eventually prove even more discriminating. At least one virus in each serogroup and 12 of 16 ungrouped rhabdoviruses are associated with arthropods either in nature or in the laboratory. Are rhabdoviruses descended from a virus of arthropods? The laboratory investigator who thinks in terms of arthropods may be led to unconventional but salubrious techniques for improved isolation and adaptation of rhabdoviruses to laboratory conditions.[32]

ACKNOWLEDGMENTS

Studies conducted in our laboratory and reported herein were supported by U.S. Public Health Service Grants AI 10984, 5 T32 AI 07098, U.S. Army Medical Research and Development Command Contract DADA17-72-C-2170, U.S. Office of Naval Research Contract N00014-78-C-0104, and the World Health Organization.

TABLE 1

Summary of Neutralization Test Results with Various Rhabdoviruses

		Mouse hyperimmune sera									
Virus	Titer-log LD_{50}	VS-Indiana	Cocal	VS-New Jersey	Piry	Chandipura	Hart Park	Flanders	Rabies	Kern Canyon	Mt. Elgon
VS-Indiana	7.1	≥6.2[a]	4.2	1.9	0	2.6	0	0	0	0	0
Cocal	7.7	4.2	≥7.0	2.7	1.7	1.8	0	0	0	0	0
VS-New Jersey	5.5	2.2	0	≥5.0	0	0	0	0	0	0	
Piry	7.9	2.3	2.6	2.5	5.7	3.4	0	0	0	0	0
Chandipura	7.3	2.5	2.7	0	4.6	≥5.7	0	0	0	0	
Hart Park	4.6	0	0	0	0	0	≥4.1	4.1	0	0	0
Flanders	4.4	0	0	0	0	0	≥3.9	≥3.9	0	0	0
Rabies	5.9	0	0	0	0	0	0	0	≥5.4	0	0
Kern Canyon	5.4	0	0	0	0	2.2	0	0	0	≥4.9	0
Mt. Elgon	5.8	0	0	0	0	0	0	0	0	0	≥5.3

[a] Log neutralization index; 0 = ≤1.5.

From Shope, R. E., *The Natural History of Rabies*, Baer, G. M., Ed., Academic Press, New York, 1973, chap. 8. With permission.

TABLE 2

Results of Plaque Reduction Neutralization Tests With Isfahan Virus and Five Vesicular Stomatitis Virus Serotypes

	Virus					
Immune serum	New Jersey	Indiana	Cocal	Piry	Chandipura	Isfahan
New Jersey	*10,240*[a]	<10	<10	<10	<10	<10
Indiana	<10	*327,680*	320	<10	<10	<10
Cocal	<10	160	*5,120*	<10	<10	<10
Alagoas	<10	20	20	<10	<10	<10
Piry	<10	<10	<10	*163,840*	80	<10
Chandipura	<10	<10	<10	<10	*10,240*	<10
Isfahan	<10	<10	40	<10	<10	*163,840*

[a] Reciprocal of highest serum dilution producing ≥95% plaque inhibition.

From Tesh, R., Saidi, S., Javadian, E., Loh, P., and Nadim, A., *Am. J. Trop. Med. Hyg.*, 26, 299, 1977. With permission.

TABLE 3

Complement-fixation Reactions of Vesiculoviruses

		Antibody					
Serocomplex	Antigen	VSV Indiana	Cocal	VSV New Jersey	Piry	Chandipura	Isfahan
VSV Indiana	VSV Indiana	256[a]	4	0	0	0	0
	Cocal	8	64	0	0	0	0
VSV New Jersey	VSV New Jersey	0	0	32	0	0	0
Piry	Piry	0	0	0	128	16	8
	Chandipura	0	0	0	4	256	8
	Isfahan	0	0	0	0	4	256

[a] Reciprocal of hyperimmune mouse ascitic fluid titer at optimal antigen dilution in grid titrations; 2 units C′; overnight incubation; 0 = <4

TABLE 4

Results of Complement-Fixation Tests Using Constant Antigen and Variable Serum Dilutions in the Examination of the Cocal, Argentina, and Brazil Strains of Vesicular Stomatitis Virus

	Virus and year of isolation							
Serum	Cocal 1961	Argentina (Cocal) 1963	Brazil Alagoas 1964	Indiana USA 1959	Indiana Mexico 1952	Indiana Venezuela 1962	Indiana Ecuador 1961	New Jersey Costa Rica 1961
Cocal	1/200[a]	1/250	1/10	1/25	1/20	<1/10	<1/10	<1/10
Argentina (Cocal)	1/180	1/100	<1/10	1/10	1/15	1/10	<1/10	<1/10
Alagoas	<1/10	<1/15	1/320	1/10	1/15	1/10	<1/10	<1/10
Indiana, USA	1/10	1/10	1/10	1/350	1/200	1/400	1/300	<1/10
Indiana, Mexico	1/15	1/10	1/40	1/80	1/120	1/200	1/160	<1/10
Indiana, Venezuela	<1/10	<1/10	<1/10	1/40	1/40	1/280	1/120	<1/10
Indiana, Ecuador	1/15	1/15	1/10	1/180	1/150	1/250	1/300	<1/10

[a] Serum dilution endpoint.

From Federer, K. E., Burrows, R., and Brooksby, J. B., *Vet. Res. Sci.*, 8, 103, 1967. With permission.

TABLE 5

Fluorescent Focus Inhibition Test Cross-reactions Among Vesicular Stomatitis Group Viruses and Absence of Cross-reactions with Other Rhabdoviruses

	Antibody						
Virus	VS-Indiana	Cocal	VS-New Jersey	Piry	Chandipura	Isfahan	Other rhabdoviruses
VS-Indiana	99[a]	97	27	0	42	51	<40
Cocal	100	100	2	6	76	98	<40
VS-New Jersey	84	0	100	46	92	40	<44
Piry	37	0	31	100	93	80	<50
Chandipura	0	0	18	100	100	100	<42
Isfahan	17	84	68	97	93	100	<50

[a] Percent reduction in fluorescent foci; hyperimmune mouse ascitic fluids reacted 1:10 with 25 to 220 fluorescent focus units of virus; CER cell monolayers; 16 to 18 hr secondary incubation.

[b] Rabies, Mokola, Lagos bat, Duvenhage, kotonkan, Obodhiang, Hart Park, Flanders, Gray Lodge, Kwatta, BeAn 157575, Mossuril, Kamese, Timbo, Chaco, Marco, Sawgrass, New Minto, Barur, Joinjakaka, Kern Canyon, Klamath, Mount Elgon bat, Navarro, Oita 296, Mosqueiro, S-1643, Yata, Keuraliba, Parry Creek, Kimberly.

TABLE 6

Neutralization[a] of Rabies Virus by Mouse Hyperimmune Ascitic Fluid to Rabies and to Four Other Related Viruses

Hyperimmune mouse ascitic fluid to	Log neutralization index
Rabies	≥5.3
Mokola	3.1
Lagos bat	
Fluid #1	0.4
Fluid #2	4.1
Obodhiang	0.3
Kotonkan	1.4

[a] i.c., infant mice; constant-serum, varying-virus technique.

From Shope, R. E., in *The Natural History of Rabies*, Baer, G. M., Ed., Academic Press, New York, 1973, chap. 8. With permission.

TABLE 7

Mouse Neutralization Test Reactions of Rabies and Duvenhage Viruses Using Undiluted Antibody[a]

	Viruses (suckling mouse brain stock)			
	Rabies		Duvenhage	
Mouse immune ascitic fluid	CVS	Street	A[b]	B
Rabies (CVS	2.0 (5.3)[c]	≤0.5 (≥2.8)	4.7 (2.4)	3.9 (2.4)
Duvenhage A	2.8 (4.5)	≤0.5 (≥2.8)	≤0.8 (6.3)	≤0.5 (≥5.8)
Duvenhage B	2.0 (5.3)	≤0.5 (≥2.8)	≤1.5 (≥5.6)	≤0.8 (≥5.5)
Titre in normal mouse ascitic fluid	7.3	3.3	7.1	6.3
Titre in diluent prior to incubation[d]	8.5	3.6	7.5	6.8

[a] In suckling mice per 0.03 m*l* i.c.; virus dilution-undiluted antibody.

[b] A and B represent different passage levels in replicate titrations.

[c] (), log neutralization index (Kärber, 1931).

[d] Diluent was 0.75% bovine albumin in phosphate-buffered saline.

From Tignor G. H., Murphy, F. A., Clark, H. R., Shope, R. E., Madore, P., Bauer, S. P., Buckley, S. M., and Meredith, C. D., *J. Gen. Virol.*, 37, 595, 1977. With permission.

TABLE 8

Serum-Dilution Neutralization Test of Rabies-Related Viruses

Virus	Inoculum: mouse LD_{50}	Serum (S) or mouse ascitic fluid (MAF): Rabies burro S	Lagos bat MAF	Mokola MAF	Control MAF (M-1056)
Rabies, strain CVS	160	15,625[a]	11	0[b]	0
Rabies, vampire bat	200	15,625	11	1	0
Lagos bat	50	5	280	18	0
Mokola	100	3	15	320	0
Control (M-1056)	125	0	0	0	25

[a] Reciprocal of 50% neutralization endpoint of serum or ascitic fluid; fivefold serum dilutions versus constant virus dilution.
[b] No survivors.

From Shope, R. E., Murphy, F. A., Harrison, A. K., Causey, O. R., Kemp, G. E., Simpson, D. I. H., and Moore, D. L., *J. Virol.*, 6, 690, 1970. With permission.

TABLE 9

Mouse Neutralization Test Reactions of Rabies and Duvenhage Viruses by Antibody Dilution Technique[a]

Mouse immune ascitic fluid	Viruses (suckling mouse brain stock): Rabies		Duvenhage	
	CVS	Street	A[b]	B
Rabies (CVS)	3,125[c]	1,325	7	9
Duvenhage A	1,420	5,500	27,000	15,000
Duvenhage B	342	940	3,125	3,840
Inoculum LD_{50}	113	75	38	50

[a] In suckling mice per 0.03 mℓ i.c. using the virus constant-antibody dilution technique; dilutions were fivefold; reagents were prepared at CDC.
[b] A and B represent different passage levels in replicate titrations.
[c] Reciprocal of 50% ascitic fluid dilution endpoint.

From Tignor, G. H., Murphy, F. A., Clark, H. F., Shope, R. E., Madore, P., Bauer, S. P., Buckley, S. M., and Meredith, C. D., *J. Gen. Virol.*, 37, 595, 1977. With permission.

TABLE 10

Cross-Plaque Neutralization Tests with Cloned Rabies, Duvenhage, Mokola, Obodhiang, and Kotonkan Viruses and Respective 50% Plaque Reduction Titers

	Titer of hyperimmune serum[a]				
Cloned virus	Rabies, CVS	Duvenhage	Mokola	Obodhiang	Kotonkan
Rabies, CVS	65,536	8,192	64	<8	<8
Duvenhage	4,096	16,384	8	<8	<8
Mokola	512	128	8,192	<8	<8
Obodhiang	<8	<8	8	8,192	8
Kotonkan	<8	<8	<8	8	512

[a] Titers expressed as reciprocals of serum dilution that give a 50% reduction of plaques.

From Buckley, S. M., *Ann. N. Y. Acad. Sci.*, 266, 241, 1975. With permission.

TABLE 11

Vaccination and Challenge of Mice with Rabies Serogroup Viruses[18]

	Vaccine[a]				
	Rabies				
Challenge[b]	HEP	HEP	TR843	Mokola	Lagos bat
Rabies, Pasteur	2.4[c]	5.0	>3.2	0.3	0.4[d]
Log LD_{50} challenge	(4.8)	(6.0)	(>6.0)	(5.5)	(6.0)
Mokola	0.8	2.2	0.1	1.9	0.6
Log LD_{50} challenge	(3.5)	(5.3)	(4.3)	(4.5)	(4.6)
Lagos bat	Not done	64.0	2.1	0	3.6
Log LD_{50} challenge		(4.5)	(4.1)	(2.3)	(4.1)

[a] HEP vaccine single intracerebral inoculation; Mokola single intraperitoneal inoculation; TR5843 and Lagos bat inoculations intraperitoneal on days 0 and 14. Mice were 3 to 6 weeks old.

[b] Intracerebral.

[c] Log protective index.

[d] Approximate value.

TABLE 12

Immunization and Challenge of Mice with Duvenhage and Strains of Rabies Virus

Group[a]	Vaccination virus	Challenge virus	Log LD_{50} of challenge virus	Log protection index
A	Rabies (HEP)	Duvenhage	3.4	1.5
	None	Duvenhage	4.9	
B	Rabies (HEP)	Rabies (Pasteur)	≤0.5	≥5.9
	None	Rabies (Pasteur)	6.4	
C	Rabies (HEP)	Maryland bat (rabies)	≤0.5	≥6.0
	None	Maryland bat	≥6.5	
D	Rabies (HEP)	Trinidad monkey TR5843 (rabies)	≤0.5	≥5.2
	None	Trinidad monkey TR5843	5.7	
E	Duvenhage	Rabies (CVS)	4.7	0.2
	None	Rabies (CVS)	4.9	
F	Duvenhage	Duvenhage	≤0.5	≥4.4
	None	Duvenhage	4.9	

[a] Groups A to D, 6-week-old mice immunized i.c. (0.03 mℓ and challenged 30 days later i.c. with serial tenfold dilutions of challenge virus, 6 mice per dilution. Groups E and F, 6-week-old mice immunized i.p. (Habel Test) and challenged i.c.

From Tignor, G. H., Murphy, F. A., Clark, H. F., Shope, R. E., Madore, P., Bauer, S. P., Buckley, S. M., and Meredith, C. D., *J. Gen. Virol.*, 37, 595, 1977. With permission.

TABLE 13

Complement-fixation Reactions of Rabies Serogroup Viruses

		Antibody					
Serocomplex	Antigen	Rabies HEP	Mokola	Lagos bat	Duvenhage	Obodhiang	Kotonkan
Rabies	Rabies pasteur	1024[a]	512	4	4	0	0
	Rabies TR5843	1024	512	64	16	0	0
	Mokola	64	512	64	0	64	0
	Lagos bat	64	512	1024	64	16	0
	Duvenhage	256	128	64	64	0	0
Obodhiang	Obodhiang	0	512	0	0	256	0
Kotonkan	Kotonkan	0	0	0	0	0	64

[a] Reciprocal of hyperimmune mouse ascitic fluid titer at optimal antigen dilution in grid titrations; 2 units C'; overnight inubation; 0 = <4

TABLE 14

Cross Complement-fixation Tests with Kotonkan and Other Viruses, Including Members of the Rabies Serogroup

	Hyperimmune mouse ascitic fluid				
Antigen	Kotonkan	Mokola	Rabies, CVS	Lagos bat	Obodhiang
Kotonkan, IbAr 23380	64[a]	32	0	0	0
Mokola, IbAn 27377	0	256	32	4	16
Rabies, CVS	0	64	256	0	0
Lagos bat	0	128	16	32	8
Obodhiang, SudAr 1154-64	0	4	0	0	128
191 arboviruses and other viruses of vertebrates	0	0	0	0	0

[a] Reciprocal of ascitic fluid titer; 0 = no reaction at ascitic fluid dilution 1:4, lowest used.

From Kemp, G. E., Lee, V. H., Moore, D. L., Shope, R. E., Causey, O. R., and Murphy, F. A., *Am. J. Epidemiol.*, 98, 43, 1973. With permission.

TABLE 15

Direct Immunofluorescence Cross-reactions of Kotonkan and Obodhiang Viruses with Members of the Rabies Serogroup (Excess Concentrations of Conjugates)

	Viruses (46 hr of infection in BHK-21 cells[a])					
	Rabies					
Conjugate	CVS strain	ERA strain	Mokola	Lagos bat	Kotonkan	Obodhiang[b]
Rabies (CVS)	4[c]	4	2	2	0	0
Mokola	2	2	4	3	1	1
Lagos bat	0	0	3	4	0	0
Kotonkan	0	0	0	0	4	0
Obodhiang	0	0	1	0	0	4

[a] Substrates shown to be 80 to 100% infected via homologous direct immunofluorescence.

[b] Obodhiang virus in Vero cells. Conjugate dilutions were used in amounts greater than fourfold excess over 4+ brightness end point dilutions (intensity scale of 4+ to nil).

[c] Intensity scale of 4+ to nil. Conjugates were used at dilution amounts greater than fourfold excess over 4+ brightness end point dilutions.

From Bauer, S. P. and Murphy, F. A., *Infect. Immun.*, 12, 1157, 1975. With permission.

TABLE 16

Direct Immunofluorescence Reactions of Duvenhage and Other Rabies Serogroup Viruses

	Antigen (suckling mouse brain impressions)				
Conjugates[a]	Duvenhage	CVS-rabies	Street rabies	Mokola	Lagos bat
Duvenhage	*3*	tr	0	0	0
CVS-rabies	2	*4*	*4*	2	1
Mokola	2	2	1	*4*	3
Lagos bat	4	tr	0	3	*4*

[a] Conjugates diluted twofold less than terminal dilution (terminal dilution defined as last dilution giving maximum brightness). Preparations were coded for reading, and graded for brightness on a scale of 0 to 4.

From Tignor, G. H., Murphy, F. A., Clark, H. F., Shope, R. E., Madore, P., Bauer, S. P., Buckley, S. M., and Meredith, C. D., *J. Gen. Virol.*, 37, 595, 1977. With permission.

TABLE 17

Indirect Immunofluorescence Cross-reactions of Kotonkan and Obodhiang Viruses with Members of the Rabies Serogroup

	Viruses (suckling mouse brain impressions[a])					
	Rabies					
Immune ascitic fluid	CVS strain	Arctic fox[b]	Mokola	Lagos bat	Kotonkan	Obodhiang
Rabies (CVS)	*1024*[c]	*512*	128	16	0[d]	0
Mokola	128	256	*1024*	512	16	64
Lagos bat	16	32	128	*128*	0	0
Kotonkan	0	0	16	0	*256*	16
Obodhiang	0	0	64	0	0	*1024*

[a] Substrates shown to be heavily infected via homologous direct immunofluorescence.
[b] An arctic fox street rabies isolate in second mouse brain passage.
[c] Ascitic fluid titer. The end point was determined as that dilution where immunofluorescence diminished from ≥3+ to 2+ or less (intensity scale of 4+ to nil).
[d] 0, No reaction at 1:8 ascitic fluid dilution.

From Bauer, S. P. and Murphy, F. A., *Infect. Immun.*, 12, 1157, 1975. With permission.

TABLE 18

Homologous and Heterologous Hemagglutination Inhibition Titers of Rabies and Some Other Bullet-Shaped Viruses

	Viruses				
Antiserum	Rabies	VSV-Indiana	VSV-New Jersey	Cocal	Kern Canyon
Rabies, burro no. 3[a]	320	<20	<20	<20	40
Rabies, burro no. 4[b]	320	<20	<20	<20	<20
Rabies, burro no. 5[c]	1280	<20	<20	<20	20
Rabies, goat no. 15[b]	160	<20	<20	<20	<20
Rabies, goat no. 19[c]	1280	<20	<20	<20	<20
Normal burro	<20	<20	<20	<20	<20
Normal goat	<20	<20	<20	<20	<20
VSV Indiana[d]	<20	80	<20	20	<20
VSV New Jersey[d]	<20	<20	160	<20	<20
Cocal[d]	<20	80	20	160	<20
Kern Canyon[d]	<20	<20	<20	<20	160
Flanders[d]	<20	<20	<20	<20	<20
Hart Park[d]	<20	<20	<20	<20	<20

[a] Prepared by immunizing with a commercial rabies vaccine for dogs (nervous tissue of caprine origin, inactivated.)

[b] Prepared by immunizing with BHK-21 cell culture grown virus.

[c] Prepared by immunizing with ecteola-cellulose column purified virus grown in suckling mouse brain.

[d] Hyperimmune ascitic fluid of mice inoculated with mouse brain suspension of virus.

From Halonen, P. E., Murphy, F. A., Fields, B. N., and Reese, D. R., *Proc. Soc. Exp. Biol. Med.*, 127, 1037, 1968. With permission.

TABLE 19

Radioimmunoassay with Rabies (HEP), Mokola, and Lagos Bat Viruses

	250 pg		
250-8000 pg	^{125}I-HEP-RNP	^{125}I-MOK-RNP	^{125}I-LBV-RNP
HEP-RNP	1.150	0.057	0.136
MOK-RNP	0.107	0.964	0.021
LBV-RNP	0.214	0.136	0.957

Note: Slope m (= tan α) of homologous and heterologous log standard curves (amounts for homologous competition are underlined).

From Bruns, M., Dietzschold, B., Schneider, L. G., and Cox, J. H., *J. Immunol. Meth.*, 18, 337, 1977. With permission.

TABLE 20

Complement-fixation Reactions of the Minor Serogroups of Rhabdoviruses

	Hyperimmune ascitic fluid										
	Hart Park group			Sawgrass group		Kwatta group		Mossuril group		Timbo group	
Antigen	Hart Park	Flanders	Mosqueiro[a]	Sawgrass	New Minto	Kwatta	BeAn 157575[a]	Mossuril	Kamese	Timbo	Chaco
Hart Park	512[b]	1024	0	0	0	0	0	0	0	0	0
Flanders	16	2048	0	0	0	0	0	0	0	0	0
Mosquiero	8	64	64	0	0	0	0	0	0	0	0
Sawgrass	0	0	0	512	4	0	0	0	0	0	0
New Minto	0	0	0	0	512	0	0	0	0	0	0
Kwatta	0	0	0	0	0	8192	8	0	0	0	0
BeAn 157575	0	0	0	0	0	8192	32	0	0	0	0
Mossuril	0	0	0	0	0	0	0	128	128	0	0
Kamese	0	0	0	0	0	0	0	64	256	0	0
Timbo	0	0	0	0	0	0	0	0	0	512	16
Chaco	0	0	0	0	0	0	0	0	0	8	32

[a] Mosquiero and BeAn 157575 are unpublished viruses supplied for study by F. P. Pinheiro and A. P. A. Travassos da Rosa, Belem, Brazil; their mention here is not intended to constitute priority of publication.

[b] Reciprocal of ascitic fluid titer with optimal antigen dilution; 2 units C′; overnight incubation; 0 = <4.

TABLE 21

Ungrouped Rhabdoviruses[a]

Barur	Oita 296[a]
Joinjakaka	Gray Lodge[a]
Kern Canyon	S 1634[a]
Klamath	Yata
Marco	Kununurra[a]
Mount Elgon Bat	Kimberly[a]
Navarro	Keuraliba

[a] Oita 296 was submitted for study by A. Oya, Gray Lodge by R. Scrivani, S 1643 by D.I.H. Simpson and Kununurra and Kimberly by N. Stanley. Their inclusion in this Table is not intended to constitute priority of publication.

TABLE 22

Cross-neutralization Tests with Five Rhabdoviruses of Fish

	Virus[a]				
Antiserum	SVC	SBI	PFD	IHN	VHS
SVC	6500	6200	20	20	<10
SBI	980	980	68	50	<10
PFD	33	28	200	27	<10
IHN	25	25	27	1700	<10
VHS	<10	<10	<10	<10	2300

[a] Reciprocal of antiserum dilutions giving 50% plaque reduction.

From Hill, B. J., Underwood, B. O., Smale, C. J., and Brown, F., *J. Gen. Virol.*, 27, 369, 1975. With permission.

REFERENCES

1. **Casals, J.,** Immunological techniques for animal viruses, in *Methods in Virology,* Vol. 3, Maramorosch, K. and Koprowsi, H., Eds., Academic Press, New York, 1967, 113.
2. **Emmons, R. W. and Riggs, J. L.,** Applications of immunofluorescence to diagnosis of viral infections, in *Methods in Virology,* Vol. 6, Maramorosch, K. and Koprowski, H., Eds., Academic Press, New York, 1977, 1.
3. **Kaplan, M. M. and Koprowski, H.,** *Laboratory Techniques in Rabies,* 3rd ed., World Health Organization, Geneva, 1973, chap. 6, 8—11, 13, 17.
4. **Dietzschold, B., Schneider, L. G., and Cox, J. H.,** Serological characterization of the three major proteins of vesicular stomatitis virus, *J. Virol.,* 14, 1, 1974.
5. **Cartwright, B. and Brown, F.,** Serological relationships between different strains of vesicular stomatitis virus, *J. Gen. Virol.,* 16, 391, 1972.
6. **Schneider, L. G., Dietzschold, B., Dierks, R. E., Matthaeus, W., Enzmann, P. -J., and Strohmaier, K.,** Rabies group-specific ribonucleoprotein antigen and a test system for grouping and typing of rhabdoviruses, *J. Virol.,* 11, 748, 1973.
7. **Shope, R. E.,** Rabies virus antigenic relationships, in *The Natural History of Rabies,* Baer, G. M., Ed., Academic Press, New York, 1973, chap. 8.
8. **Cartwright, B. and Brown, F.,** Serological relationships between different strains of vesicular stomatitis virus, *J. Gen. Virol.,* 16, 391, 1972.
9. **Federer, K. E., Burrows, R., and Brooksby, J. B.,** Vesicular stomatitis virus — the relationship between some strains of the Indiana serotype, *Vet. Res. Sci.,* 8, 103, 1967.
10. **Tesh, R., Saidi, S., Javadian, E., Loh, P., and Nadim, A.,** Isfahan virus, a new vesiculovirus infecting humans, gerbils, and sandflies in Iran, *Am. J. Trop. Med. Hyg.,* 26, 299, 1977.
11. **Murphy, F. A. and Fields, B. N.,** Kern Canyon virus: electron microscopic and immunological studies, *Virology,* 33, 625, 1967.
12. **Halonen, P. E., Murphy, F. A., Fields, B. N., and Reese, D. R.,** Hemagglutination of rabies and some other bullet-shaped viruses, *Proc. Soc. Exp. Biol. Med.,* 127, 1037, 1968.
13. **Myers, W. L. and Hanson, R. P.,** Immunodiffusion studies on the antigenic relationships within and between serotypes of vesicular stomatitis virus, *Am. J. Vet. Res.,* 23, 896, 1962.
14. **Reichman, M. E., Schnitzlein, W. M., Bishop, D. H. L., Lazzarini, R. A., Beatrice, S. T., and Wagner, R. R.,** Classification of the New Jersey serotype of vesicular stomatitis virus into two subtypes, *J. Virol.,* 25, 446, 1978.
15. **Buckley, S. M.,** Arbovirus infection of vertebrate and insect cell cultures, with special emphasis on Mokola, Obodhiang, and kotonkan viruses of the rabies serogroup, *Ann. N. Y. Acad. Sci.,* 266, 241, 1975.
16. **Tignor, G. H., Murphy, F. A., Clark, H. F., Shope, R. E., Madore, P., Bauer, S. P., Buckley, S. M., and Meredith, C. D.,** Duvenhage virus: morphological, biochemical, histopathological, and antigenic relationships to the rabies serogroup, *J. Gen. Virol.,* 37, 595, 1977.
17. **Shope, R. E., Murphy, F. A., Harrison, A. K., Causey, O. R., Kemp, G. E., Simpson, D. I. H., and Moore, D. L.,** Two African viruses serologically and morphologically related to rabies virus, *J. Virol.,* 6, 690, 1970.
18. **Tignor, G. H. and Shope, R. E.,** Vaccination and challenge of mice with viruses of the rabies serogroup, *J. Infect. Dis.,* 125, 322, 1972.
19. **Kemp, G. E., Lee, V. H., Moore, D. L., Shope, R. E., Causey, O. R., and Murphy, F. A.,** Kotonkan, a new rhabdovirus related to Mokola virus of the rabies serogroup, *Am. J. Epidemiol.,* 98, 43, 1973.
20. **Bauer, S. P. and Murphy, F. A.,** Relationship of two arthropod-borne rhabdoviruses (kotonkan and Obodhiang) to the rabies serogroup, *Infect. Immun.,* 12, 1157, 1975.
21. **Burns, M., Dietzschold, B., Schneider, L. G., and Cox, J. H.,** Comparison of the ribonucleoproteins of different rabies virus serotypes by radioimmunoassay, *J. Immunol. Meth.,* 18, 337, 1977.
22. **Boyd, K. R.,** Serological comparisons among Hart Park virus and strains of Flanders virus, *Infect. Immun.,* 5, 933, 1972.
23. **Berge, T. O., Ed.,** International Catalogue of Arboviruses, 2nd ed., Department of Health, Education, and Welfare Publication No. (CDC) 75-8301, 1975, 364.
24. **Ritter, D. G., Calisher, C. H., Muth, D. J., Shope, R. E., Murphy, F. A., and Whitfield, S. G.,** New Minto virus: a new rhabdovirus from ticks in Alaska, *Can. J. Microbiol.,* 24, 422, 1978.
25. **Monath, T. P., Cropp, C. B., Frazier, C. L., Murphy, F. A., and Whitfield, S. G.,** Viruses isolated from reptiles: identification of three new members of the family Rhabdoviridae, *Arch. Virol.,* 60, 1, 1979.
26. **Karabatsos, N., Ed.,** Supplement to international catalogue of arboviruses, *Am. J. Trop. Med. Hyg.,* 27, 380, 1978.

27. **Hill, B. J., Underwood, B. O., Smale, C. J., and Brown, F.**, Physico-chemical and serological characterization of five rhabdoviruses infecting fish, *J. Gen. Virol.*, 27, 369, 1975.
28. **McCain, B. B., Fryer, J. L., and Pilcher, K. S.**, Antigenic relationships in a group of three viruses of salmonid fish by cross neutralization, *Proc. Soc. Exp. Biol. Med.*, 137, 1042, 1971.
29. **Jorgensen, P. E. V.**, Egtved virus: antigenic variation in 76 virus isolates examined in neutralization tests and by means of the fluorescent antibody technique, *Symp. Zool. Soc. London*, 30, 333, 1972.
30. **Ahne, W.**, A rhabdovirus isolated from grass carp (Ctenopharyngodon idella Val.), *Arch. Virol.*, 48, 181, 1975.
31. **Sokol, F., Clark, H. F., Wiktor, T. J., McFalls, M., Bishop, D. H. L., and Obijeski, J. F.**, Structural phosphoproteins associated with ten rhabdoviruses, *J. Gen. Virol.*, 24, 433, 1974.
32. **Buckley, S. M.**, Singh's *Aedes albopictus* cell cultures as helper cells for the adaptation of Obodhiang and kotonkan viruses of the rabies serogroup to some vertebrate cell cultures, *Appl. Microbiol.*, 25, 695, 1973.

Chapter 4

ELECTRON MICROSCOPY OF THE RHABDOVIRUSES OF ANIMALS

Frederick A. Murphy and Alyne K. Harrison

TABLE OF CONTENTS

I. INTRODUCTION

The term "rhabdovirus" emerged after it was found that the unusual bullet-shaped morphology of vesicular stomatitis virus was shared with other viruses, such as rabies and sigma viruses. Melnick and McCombs[1] first suggested this term in 1966, and the International Committee on Nomenclature of Viruses recommended it in 1970[2] after the term "stomatovirus" was rejected as inappropriate. In the present universal taxonomic scheme of the International Committee on Taxonomy of Viruses,[3] usage has been formalized as the family Rhabdoviridae. The first morphologic studies leading to these taxonomic decisions were those of Chow et al.[4] and Bradish et al.,[5] who established that the infectious particle of vesicular stomatitis virus was rod shaped. Subsequently, more in-depth studies showed that this virus had unique structural details and an overall bullet-shaped outline.[6-8] Similar structure was found for rabies virus,[9-13] and in succeeding years large numbers of morphologically similar viruses have been isolated and characterized from sources as diverse as fish, plants, and arthropods. Reviews by Howatson,[14] Knudson,[15] Murphy,[16,17] and Bishop and Smith[18] contain descriptions of most of these viruses.*

Perhaps more than any other major taxonomic grouping of viruses, the family Rhabdoviridae has been constructed from the morphologic and morphogenetic characteristics of member viruses; that is, electron microscopic examination leaves little doubt whether a new candidate virus is a rhabdovirus or not, and without electron microscopic study few new viruses would be removed from an "uncharacterized" category of RNA, solvent-sensitive viruses. Such has been the case with several rhabdoviruses considered "ungrouped arboviruses" before electron microscopic examination.[19] The few serologic interrelationships among animal rhabdoviruses have not allowed much subgrouping (exception, *Lyssavirus* genus), and comprehensive comparisons of physicochemical properties (RNA and proteins) have not yet been made. This overreliance upon electron microscopic study has resulted in slow progress in our understanding of the family as a whole.

Despite the massive amount of information available on the constituency and replication of vesicular stomatitis virus,[18,20] and to a much lesser degree of rabies virus,[21-23] little work has been done on most animal rhabdoviruses — with many of the newly recognized viruses there has been no nonmorphologic study.

In this chapter morphologic and morphogenetic characterizations are presented for the virus family as a whole, and then these characterizations are used for informal subgrouping of the family. In most cases, these subgroupings result in only vague clustering of serologically independent viruses, or small virus serogroups; only in the *Vesiculovirus* and *Lyssavirus* genera does the morphologic clustering correspond to formalized subgrouping based upon close similarities of multiple virion properties. If in the case of the poorly characterized rhabdoviruses these morphologic clusterings have "predictive" value for anticipated comparative physicochemical studies, then the family substructuring is worthwhile as an interim exercise. On the other hand, if in considering the varying ecologic sources of the known rhabdoviruses of animals it becomes clear that the evolutionary diversion into unique econiches has gone so far that clusterings are meaningless, then other, nonmorphologic, bases for determining relationships in the virus family will have to be sought.

* Listing of previously undescribed viruses in this chapter is not intended to constitute priority description.

II. MORPHOLOGY OF RHABDOVIRUSES

A. Negative Contrast Electron Microscopy

All vertebrate rhabdovirus particles are bullet-shaped, that is, they are either cylindrical or cone-shaped, with one end hemispherical and the other planar or concave. The major constituents of the virus particles from their surface inward are (1) a surface projection layer, (2) a unit-membrane envelope layer, and (3) a helically wound ribonucleocapsid[17,24-26] (Figures 1 to 4)*. Each of these constituents will be described in detail.

The surface projection layer consists of clearly discernible spikes set in the envelope surface at 4 to 5 nm intervals. It has been estimated that the surface of a vesicular stomatitis virus particle has 500 spikes.[27] In most preparations the individual spikes appear randomly spaced, but in some cases, especially with rabies virus, there is sixfold or honeycomb symmetry in the placement of spikes (Figures 2 and 5). This ordered placement may reflect symmetry within the envelope layer and in the morphologically indistinct matrix layer immediately beneath it.[27,28] The length of the spikes, equivalent to the thickness of the outer layer, has ranged between 6 and 10 nm in various descriptions; because the same wide range has been reported for the same virus (e.g., vesicular stomatitis virus), it is probable that there is a uniform spike lengh for all animal rhabdoviruses and that variations reflect differences in methods of preparation and means of measuring. Individual spikes often appear to have a knoblike distal configuration with a narrow stalk, but such detail is not always resolved.[16,26,29,30] As a result of this uncertainty, various models of rhabdoviruses alternately depict spikes as rod-shaped structures.[14,16,18,20,22,27,28,31] Because most rhabdovirus models have been designed primarily to schematically place the chemically determined number of constituent subunits, analogy to particles observed by electron microscopy is often inaccurate. Such "constituent" models appear properly elsewhere in this volume. The model illustrated in Figure 3, on the other hand, represents an idealized rhabdovirus particle as seen with negative contrast electron microscopy. The character and means of insertion of spikes are purposely left unresolved, since this matter has not been clearly depicted by electron microscopy per se.[27,29,30]

The envelope layer is clearly composed of a lipoprotein unit membrane derived from host cell membrane.[32] It is asymmetric in that its outer face is covered with spikes and its inner face is covered with a matrix protein layer. The fluid nature of the envelope layer may be overlooked in some negative contrast and thin-section micrographs of rhabdoviruses, but it is confirmed by the close apposition of envelope to nucleocapsid (even conforming to defects or breaks in nucleocapsid), and especially by the invagination of envelope into the hollow planar end of the nucleocapsid to form a stain-filled axial channel (Figures 1 to 4). This fluidity or plasticity of the rhabdovirus envelope is also evidenced by the common presence of "terminal blebs". These membranous structures are formed at the planar end of many virus particles by the avulsion of excess host cell membrane at the site of budding and pinching off (Figure 3). The thickness of the viral envelope as seen in negative contrast, is not indicative of its native state; its unstructured electron lucent appearance is artifactually thickened (Figures 1 and 4).

The matrix layer is usually not discerned in the electron dense, narrow space between the lucent shells representing virus envelope and nucleocapsid. However, Cartwright et al.,[27] in their vesicular stomatitis model, concluded that a hexagonal lattice arrangement of matrix protein subunits is most likely. Neurath et al.[30] and Vernon et al.[28] found that a similar lattice best fit their rabies virus model.

* All figures appear following text.

The nucleocapsid is revealed when negative contrast medium penetrates the viral envelope layer (Figures 1 and 4). The helically coiled structure of the nucleocapsid appears as cross striations which have an interval of 4.5 nm[14,16,26,33] (Figure 1). When the intact nucleocapsid is freed from its surrounding envelope, either spontaneously via storage at 4°C or with detergent treatment, it is resolved as a precisely wound single helix which, in vesicular stomatitis viruses, is formed from a nucleoprotein strand 3.6 μm long with approximatly 30 turns of equal diameter and four or five turns of diminishing diameter at the rounded end. Of all the models of rhabdovirus nucleocapsid stucture, that of Nakai and Howatson[26] best matches high resolution negative contrast observations of intact and unwound nucleocapsid helices. In this model the nucleoprotein strand is described as a "ribbon" because each of its approximately 1000 protein units* (dimensions 9 × 3 × 3 nm) is strung with its long dimension perpendicular to the RNA core. This "ribbon-shaped' nucleoprotein strand is wound into the helix with the protein units arranged radially and the nucleic acid buried within. When the helical structure unwinds, the "ribbon" is twisted, and in many instances rewinding occurs, so that a helix is formed with a diameter approximately one half that of the native virus nucleocapsid.

The nucleoprotein strand length, the number of turns forming the complete nucleocapsid helix, and the overall nucleocapsid shape and dimensions vary among the animal rhabdoviruses. These variations fall into four arbitrary categories. First, the prototype vesicular stomatitis virus nucleocapsid has a precisely cylindrical form, a diameter of 50 nm, and a length of approximately 165 nm; nucleocapsids of the other members of the *Vesiculovirus* genus are indistinguishable[14,26,29] (Figure 1). In most observations, rabies virus and the other members of the *Lyssavirus* genus have appeared to have very similar nucleocapsid dimensions. However, the length of the rabies virus nucleoprotein strand has alternately been reported as 4.2 μm and 3.6 μm, a significant variation.[18,20,34] Similarly, even though the rabies virus nucleocapsid has usually been found to be 165 nm long with 30 to 35 turns to its helix,[16,34] it is clear that under some circumstances nucleocapsids from infectious virus particle preparations are shorter and have fewer turns[90] (Figure 2). This discrepancy requires further investigation; length measurements must be obtained from street viruses for comparison with virus strains which have been passaged in laboratories for many years, and such measurements must be made with careful distinction between infectious B-particles and defective-interfering long T-particles which may resemble infectious particles.

A second category of nucleocapsid form and dimensions includes the viruses with cylindrical nucleocapsids but shorter lengths and fewer turns, corresponding to shorter overall virus particle length. An example is Kern Canyon virus[35] (Figure 4). A third grouping includes the viruses with cylindrical nucleocapsids which are longer than those of vesiculoviruses. Mt. Elgon bat virus is representative[36] (Figure 6). Finally, a fourth grouping includes viruses with nucleocapsids that appear conical. Although the basis for this variation is not fully understood, it is clear that the conical shape is not simply a flattening artifact. For example, vesicular stomatitis nucleocapsids never show spreading of their planar end. Viruses such as kotonkan virus[37,38] (Figure 7) and bovine ephemeral fever[39-42] (Figure 8) exhibit varying broadening of their planar ends depending upon the stage of infection; early infection yields nearly cylindrical nucleocapsids, and late infection yields very broad conical forms.

T-particles, which have been observed with all animal rhabdoviruses appropriately

* Chemical determinations indicate that there are approximately 2000 to 2300 molecules of nucleoprotein per virion.

studied, clearly have the same basic construction as infectious B-particles.[43-45] Surface projections and envelope are unremarkable. The nucleocapsid of the first vesicular stomatitis virus T-particle to be studied was about 75 nm long; total particle length was 82 nm.[43] Subsequently, T-particles as short as 65 nm with as few as 8 to 10 cross striations were obtained by serial undiluted passage;[44] in other circumstances (e.g., with temperature sensitive mutants of vesicular stomatitis virus[46]) a diversity of lengths between 65 nm and the 165 length of B-particles has been obtained. In most T-particle preparations, large numbers of anomalous short particles with two flat ends are intermixed with particles with one round and one flat end.

Rabies virus T-particles have modal lengths between 70 and 100 nm. Excessively long anomalous virus particles are observed in most rhabdovirus specimens either from cell culture or infected animal tissue. These particles may be up to 400 nm long and contain an appropriately increased number of turns of their nucleocapsid helix. Other anomalies also occur; some particles contain more than one nucleocapsid within a common envelope, forming V-shaped (two capsids), Y-shaped (three capsids), or rectangular (two, three, or four capsids side-by-side) forms.

B. Thin-Section Electron Microscopy

In general, thin-section electron microscopy must be considered secondary to negative contrast methods in elucidation of viral morphology per se, but with the rhabdoviruses the techniques are complementary. Vesicular stomatitis virus particles were first described in thin sections by Reczko[6] and rabies virus by Matsumoto.[47,48] These early observations weighed heavily in the association of infectivity with the bullet-shaped particles described in shadow-cast and negative contrast preparations.

In thin section, rhabdoviruses appear to be bullet-shaped only when sectioned exactly longitudinally; they have circular or eliptical profiles when sectioned transversely or obliquely (Figures 9 to 11). Despite variations in fixation, embedding, and staining methods, there is a remarkable uniformity in the depiction of the morphology of particular rhabdovirues (e.g., vesicular stomatitis virus, rabies virus) among laboratories throughout the world.[49-53] This allows more meaningful comparisons of new rhabdoviruses to be made than is possible with negative contrast methods where methodology affects virus appearance. In all cases, the surface projection layer appears as a thick and delicate zone entirely surrounding virus particles (Figure 12). Individual spikes are not well resolved. Usually spikes do not extend beyond the junction of budding virus particles and adjacent host cell membrane; however, in instances where virtually the entire plasma membrane of an infected cell (cell culture only) is involved in virus budding, spikes can occur randomly in patches upon otherwise undifferentiated membrane.[54] The unit-membrane structure of viral envelope and its contiguity with host cell membrane are clear where budding particles are favorably oriented to the plane of section and where sections are particularly thin. Depending upon the thickness of sections, the nucleocapsid has a varying appearance. In some longitudinally sectioned particles, the turns of the nucleocapsid helix are seen clearly as cross striations. In very thin sections of particles cut through their centers, the nucleocapsid is seen as a dense beaded layer immediately beneath the envelope (Figure 12); such particles have an especially lucent center space. In the more commonly seen thicker sections, the entire nucleocapsid may appear as a solid, dense form without further detail. Some of the other morphologic features of rhabdoviruses as seen in negative contrast are confirmed in thin sections: "terminal blebs" are common; axial channels are clear in longitudinally sectioned particles and in a few transversely sectioned particles; anomalously long and branched particles are also evident.

Measurement of the dimensions of particles in thin section have been accumulated from studies in our laboratory of many of the animal rhabdoviruses in various host systems (unknown shrinkage from glutaraldehyde fixation; approximately 1% polymerization shrinkage from epoxy embedment): mean diameter of particles including surface projection layer has ben 75 nm; mean diameter excluding surface projection layer has been 60 nm; mean diameter of nucleocapsid has been 50 nm; mean diameter of interior lucent space has been 38 nm; mean thickness of surface projection layer has been 9 nm. Mean length of viruses of the *Vesiculovirus* and *Lyssavirus* genera has been 180 nm; lengths of other rhabdoviruses are described in the next section.

C. Morphologic Characterization of the Animal Rhabdoviruses

As described, many of the morphologic characteristics of the animal rhabdoviruses appear to be invariable and of primary value in defining the virus family per se. However, virus particle mean length and shape (nucleocapsid shape) are variable and have value in contributing to a scheme for subgrouping. Table 1 represents an attempt to use these morphologic characteristics together with (1) morphogenetic and ultrastructural pathologic characteristics, and (2) serologic interrelationships for an informal and arbitrary subgrouping scheme. In the table, horizontal lines distinguish subgroupings; solid lines separate the clearly defined *Vesiculovirus* and *Lyssavirus* genera, the morphologically similar long viruses (e.g., Flanders virus), and all the individual viruses which cannot be clustered because of unique characteristics or more often, lack of sufficient information. A dotted line is used to separate the lyssaviruses from kotonkan and Obodhiang viruses which have a distinctly conical particle shape, but a reproducible serologic relationship to some lyssaviruses. A double line is used to separate the fish rhabdoviruses and the invertebrate rhabdoviruses. Serogroups are designated with captal letters at the left margin, and members of serogroups are designated with numbers. The source of isolates, their geographic distribution, and the associated diseases caused in nature or experimental animals are included to more precisely identify the listed viruses. Morphologic/morphogenetic clusterings of the viruses listed in the table produce four categories, corresponding to those described in the preceding section for nucleocapsid variations. Several newly recognized viruses are included in the table which have not been characterized beyond the basic electron microscopic observation of rhabdovirus shape.

III. MORPHOGENESIS AND ULTRASTRUCTURAL ASPECTS OF INFECTION

A. Morphogenetic Aspects of Infection in Cultured Cells

The early stages of rhabdovirus infection, adsorption, and entry have been studied by electron microscopy, but results have been controversial. Two entry processes have been advanced:

1. Heine and Schnaitman[55] first claimed that the penetration of vesicular stomatitis virus occurs via fusion between viral envelope and host cell membrane with entry of free nucleocapsid directly into cytoplasm.
2. Simpson et al.[56] maintained that entry occurs via viropexis (phagocytosis) wherein virus taken into a phagocytic vacuole is degraded and nucleocapsid is delivered into cytoplasm.

Iwasaki et al.,[57] studying rabies virus infection in cultured cells, found a combination of both processes; virus was taken into phagocytic vacuoles, and then virus-fused with vacuolar membrane to deliver nucleocapsid into cytoplasm. These variances have not been further resolved; experiments directed at this matter must reflect physiologic conditions — the indiscriminately vigorous phagocytic activity of many cells in culture must not lead to a bias toward a phagocytic entry mechanim, and the vigorous capacity of vesicular stomatitis virus to cause cell fusion (even massive syncytia) in cell cultures must not lead to an unproven extension of the fusion entry mechanism to in vivo infection condition.

The assembly of rhabdovirus constituents into virus particles and the subsequent release of particles from infected cells involves coordinated processes which are remarkable in their speed and magnitude. All rhabdoviruses are assembled via a budding process occurring upon host cell membrane. Deformation of host cell membrane into the envelope layer of a nascent virus particle is coordinated with the coiling of a preformed nucleoprotein strand into a nucleocapsid structure; that is, nascent budding particles of varying length have an assembled nucleocapsid only as long as the envelope deformation itself. This suggests that the lengthening of the nucleocapsid helix directs the fluid host cell membrane into its envelope configuration. The insertion of spikes into the envelope is also coordinated with nascent nucleocapsid coiling. Spikes are present on even the shortest budding particles and extend to the base of budding particles of all lengths. Pinching off of mature virus particles occurs after nucleocapsid coiling is complete; the process often results in avulsion of extra membrane, which becomes the "terminal bleb" on virus particles. The fluid nature of membrane and envelope allows sealing off of the planar end of virus particles and of the bud site on the host cell. The rhabdovirus budding process can be extremely fast; in some cases, where virus production is proceeding rapidly, only few intermediate length particles may be seen upon membranes crowded with full length particles. Anomalous rhabdovirus particles most commonly result when the pinching-off process fails so that two or more nucleocapsids in varying orientations are contained within a common envelope.

Only one exception has been reported to the general observation that rhabdovirus maturation requires preformed host cell membrane involvement. Hummeler et al.[58] found that rabies virus particles formed in cytoplasm as a consequence of apparent de novo envelope synthesis. This observation has not been confirmed under conditions where the entire infection course has been followed. Before considering de novo envelope synthesis further, we must prove that intracytoplasmic membrane systems are not destroyed at a time after conventional virus budding has occurred but before other cytopathic changes are evident.

The predominant site of budding varies among rhabdoviruses and is influenced by the host cell system. For example, vesicular stomatitis virus buds mainly upon internal membranes in most primary and continuous cell lines, but in Vero and L cells it is associated chiefly with plasma membranes.[6,59] Rabies, on the other hand, buds indiscriminately upon plasma membrane, endoplasmic reticulum, and even the outer lamella of the nuclear envelope in most cell culture systems.[16,58,60] Iwasaki et al. have shown that the earliest rabies virus budding in BHK-21 cells is from plasma membranes, and that other sites predominate later.[57,61-63]

Rhabdovirus assembly is usually associated with the accumulation of extra nucleoprotein strands within host cell cytoplasm. Early in infection the strands may be undetectable by electron microscopy because of their narrow diameter and dispersion in cytosol. When first detectable, nucleoprotein strands are in the form of loose, delicate masses which begin to displace host cell material. Continued production leads to increased compaction and focally complete displacement of host cell organelles. In cells infected with the most rapidly growing rhabdoviruses (vesicular stomatitis viruses, for example), nucleoprotein accumulation may not reach visible density, or may only reach

Table 1
THE RHABDOVIRIDAE FAMILY

Rhabdoviruses of Vertebrates and Invertebrates[a]

Genus[b] or Serologic Group[c]			Virus Name (Species) (Serotype)	Source of[d] [e] Isolates	Geographic[e] Distribution	Disease in Nature (Experimental Animals)	Ultrastructural Characteristics: Length of Virus Particles in nm	Shape of Virus Particles[f]	In vivo Site of Budding	Inclusion[h] Bodies
A	Vesiculovirus Genus	1.	Vesicular stomatitis Indiana (prototype species of family)	man, cattle, horses, sandflies, mosquitoes	New World	vesicular disease (encephalitis, hepatitis, vesicular disease)	180	B	PM	±
		2	Cocal (VSV-Argentina)	horses, rodents, mites, mosquitoes	Trinidad, Argentina, Brazil	vesicular disease (encephalitis, hepatitis, vesicular disease)	180	B	PM	±
		3.	Alagoas (VSV-Brazil)	horses	Brazil	vesicular disease (encephalitis, hepatitis, vesicular disease)	180	B	PM	NA
		4	Vesicular stomatitis New Jersey	man, cattle, horses	New World	vesicular disease (encephalitis, hepatitis, vesicular disease)	180	B	PM	±
		5	Chandipura	man hedgehogs, sandflies	India Nigeria	(encephalitis, hepatitis, vesicular disease)	180	B	PM	±
		6.	Piry	opossum	Brazil	(encephalitis, hepatitis, vesicular disease)	180	B	PM	±
		7.	Isfahan	(man) (gerbil) sandflies	Iran (USSR)	(encephalitis, hepatitis, vesicular disease)	180	B	PM	NA
B	Lyssavirus Genus	1	Rabies	man, bats, carnivores, etc.	Worldwide with known exceptions	rabies	180	B	ER/PM	++
		2.	Lagos bat	bats	Nigeria Central African Empire	(encephalitis)	180	B	ER/PM	++
		3.	Mokola	man, shrews	Nigeria, Cameroon	rabies like disease (encephalitis)	180	B	ER/PM	++
		4.	Duvenhage	man (bat)	South Africa	rabies-like disease (encephalitis)	180	B	ER/PM	++
C		5.	Kotonkan	(cattle), culicoides	Nigeria	ephemeral fever-like disease (encephalitis)	180	C	PM	+
D		6	Obodhiang	mosquitoes	Sudan	(encephalitis)	180	C	PM	+
E.			Bovine ephemeral fever	cattle, culicoides	Japan, Australia, Africa, Iran	ephemeral fever (encephalitis)	185	C/B	PM	+
F.			Marco	lizards	Brazil	(encephalitis)	180	C	PM/ER	
G			Kern Canyon	bats	California	(encephalitis)	132	SB	PM	±
H			Klamath	rodents	California, Alaska	(encephalitis)	167	SB	ER	++
I.			Barur	rodents, bats, ticks, mosquitoes	India, Kenya	(encephalitis)	149	SB	PM	
J.		1.	Flanders	birds, mosquitoes	North America	(encephalitis)	218	LB	PM	+
		2.	Hart Park	birds, mosquitoes	North America	(encephalitis)	218	LB	PM	+
K.			Mount Elgon bat	bats	Kenya	(encephalitis)	230	LB	PM	++
L.			Joinjakaka	mosquitoes	New Guinea	(encephalitis)	197	LB	PM	++
M.			Navarro	birds	Colombia	(encephalitis)	220	LB	ER/PM	++
N.		1.	Kwatta	mosquitoes	Surinam	(encephalitis)	230	LB	ER/PM	+
		2.	BeAn 157575	birds	Brazil	NA	NA		NA	NA
O.		1.	Mossuril	birds, mosquitoes	Africa	(encephalitis)	246	LB	PM	NA
		2.	Kamese	mosquitoes	Africa	NA	NA		NA	NA
P.		1.	Sawgrass	ticks	Florida	(encephalitis)	220	LB	PM	±
		2.	New Minto	ticks	Alaska	(encephalitis)	220	LB	NA	NA
Q.			Chaco	lizards	Brazil	(encephalitis)	202	LB	ER	

R.	S-1643	mosquitoes	Sarawak	NA	NA	NA	NA	NA
S.	Grey Lodge	mosquitoes	California	NA	NA	NA	NA	NA
T.	Oita - bat	bats	Japan	NA	NA	NA	NA	NA
U.	Kimberley	mosquitoes	Australia	NA	NA	NA	NA	NA
V.	Parry Creek	mosquitoes	Australia	NA	NA	NA	NA	NA
W.	Kununurra	mosquitoes	Australia	NA	130	SB	NA	NA
X.	Yata	mosquitoes	Central African Empire	NA	185	B	NA	NA
Y.	Keuraliba	gerbils, Mastomys	Senegal	NA	195	B	NA	NA
Z.	Egtved (viral hemorrhagic septicemia)	trout	Europe	hemorrhagic septicemia	180	B	PM	NA
AA.	Infectious hematopoietic necrosis virus	salmon and trout	United States	hematopoietic necrosis	188	B	PM	NA
BB.	Spring viremia of carp	cyprinids	Europe	systemic hemorrhagic disease	180	B	PM	NA
CC.	Red disease of Pike (Pike fry rhabdovirus)	esocids	Europe	hemorrhagic septicemia hydrocephalus	160	B	PM	NA
DD.	Rhabdovirus of eels	anguillids	Japan	NA	173	B	NA	NA
EE.	Rhabdovirus of grass carp	grass carp	Europe	systemic hemorrhagic disease	180	B	PM	NA
FF.	Sigma	Drosophila		(CO_2 Sensitivity)	160	SB	PM	NA
GG.	Rhabdovirus of entamoeba	E. histolytica, E. invadens						
HH.	Rhabdovirus of blue crab	crab	Canada		115	SB	ER	NA

a. This table has been compiled from data obtained from many sources, and has been assembled in the Viral Pathology Branch, Center for Disease Control, Atlanta, Georgia, and the Yale Arbovirus Research Unit, New Haven, Connecticut. Listing of previously undescribed viruses in this table is not intended to constitute priority description. Several viruses isolated only from arthropods are listed with the viruses from vertebrates; this is done arbitrarily because such viruses are infectious, usually lethal, for newborn mice when inoculated intracerebrally.

b. Horizontal lines have been used to distinguish subgroupings. With the vesiculoviruses and lyssaviruses this distinction is at the genus level. In other cases subgroupings are informal and primarily based upon clusterings of morphologic and morphogenetic characteristics. The dotted horizontal line separates lyssaviruses from two viruses which have distinct morphologic characteristics, but a reproducible serologic relationship to lyssaviruses. The doubled horizontal lines separate the fish rhabdoviruses (which do not infect mice) and the invertebrate viruses.

c. Serogroups are designated by capital letters at the left margin. Members of serogroups are designated with numbers.

d. Parentheses indicates serologic evidence only.

e. Tabulated from Annual Report of Yale Arbovirus Research Unit, New Haven.

f. Abbreviations are B, typical rhabdovirus particle morphology bullet shaped with parallel sides and length of approximately 180nm. SB, typical morphology with parallel sides, but characteristically shorter than prototype VSV. LB, typical morphology with parallel sides, but characteristically longer than prototype VSV. C, conically shaped virus particles and conically wound nucleocapsids. NA, data not available.

g. Abbreviations are PM, plasma membrane. ER, endoplasmic reticulum (and other intracytoplasmic membranes).

h. Symbols are ++, inclusion bodies (massed nucleocapsid material) are prominent - large and widespread in infected cells. +, inclusion bodies are present in most instances, but not too prominent. ±, inclusion bodies present under some circumstances, but not characteristic of the infection. −, inclusion bodies not found.

a stage of indefinite inclusion body formation (Figures 10 and 11). In a few cell lines, vesicular stomatitis virus infection is marked by the formation of precisely marginated inclusion bodies.[33] Those rhabdoviruses which grow and cause cytopathic changes more slowly, (such as rabies viruses) usually form large inclusion bodies in cultured cells as nucleoprotein accumulation is condensed in precisely marginated areas of cytoplasm.[16] These nucleoprotein inclusions are seen as smoothly eosinophilic bodies by light microscopy and as smoothly brilliant bodies when appropriately stained and examined by immunofluorescence microscopy. Because many rhabdoviruses have not been propagated in cell culture, the relative propensity for formation of nucleoprotein inclusions is better catalogued from examination of tissues from infected animals. This is described in the following section.

As a result of many rhabdovirus infections, cultured cells undergo progressive cytopathic changes, starting with rounding and swelling and proceeding to vacuolation, rarifaction or condensation, and ultimately to frank disruption of membranes. Such changes usually occur late in relation to peak particle production, whether this peak be at 3 to 4 hr, as with vesicular stomatitis viruses, or at several days, as with fixed rabies viruses. In other instances, rhabdoviruses (such as street rabies virus strains) are known to cause very little cytopathology even at the ultrastructural level of observation, but studies of such viruses in cell culture systems have not been extensive.

B. Morphogenetic Aspects of In Vivo Infections

The single common substrate for propagating all of the animal rhabdoviruses is the brain of suckling mice. Many of the viruses were initially isolated by intracerebral inoculation of suckling mice and were quantitated by tritration of lethality endpoints. With some of the rhabdoviruses, propagation has only been attempted in mouse brain, and with several others, adaptation to cell culture has been difficult or impossible. The overall character of rhabdovirus infections in the neurons of suckling mice is similar to that in cell cultures. No variations in particle dimensions have been noted with those viruses studied comparatively in cell culture and in mouse brain. In most instances there is a high degree of uniformity in virus particle length, shape, and general morphologic appearance in in vivo infection sites. Therefore, ''ultrastructural characteristics'' of those viruses which have not been studied in cell cultures have been taken from in vivo studies for inclusion in Table 1. In most cases, particle length values represent means of 100 measurements of free virus particles lying longitudinally in the plane of section.

As in cell culture, virus assembly involves budding from host cell membranes, but budding sites are even more restricted in animal tissue. For example, it was believed for some years that street rabies virus strains budded only from endoplasmic reticulum membranes of neurons,[16,21,64] whereas fixed virus strains also budded upon plasma membranes.[16] A similar lack of street virus budding upon plasma membranes was shown in fox brain[65] and in neuronal organ culture.[66,67] These findings led to a problem in explaining the means for street virus dissemination in brain; the matter has now been explained in studies with other lyssaviruses[23,68,69] (Mokola, Lagos bat viruses) and, in convincing detail, by Iwasaki et al.[61,62] in studies with street and fixed rabies virus strains. In these latter studies, it was shown that, terminally, most virus is trapped within neurons after budding upon internal membranes, but that early in the rabies infection course a high proportion of virus is formed upon plasma membranes and is free to disseminate infection via spread through intercellular spaces. This change in membrane system preference at different stages of infection matches that of rabies virus infections in cell cultures.[57] Membrane system preference also varies between different host organs and tissues. In contrast to the dissemination described in the brain,[68-70] rabies virus buds in salivary gland primarily upon plasma membranes of mucous cells facing the salivary secretion space; this delivers high titers of infectious virus into the saliva for bite transmission and is essential for the survival of the virus in nature.[17,65] In other rhabdovirus infections of the organs of experimental animals, membrane preference is precise and favorable to the spread of infection. For example, in the primary target organs of vesicular stomatitis virus infection of mice and hamsters (the liver and kidneys), only plasma membrane budding is seen (Figure 9).[71] This yields the viremia necessary for transmission via bloodsucking arthopods.

Inclusion bodies have been studied in rhabdovirus infections of animals for many years. The Negri body has long been considered pathognomonic of rabies virus infection and still has diagnostic value where more sophisticated diagnostic methods are not available. In 1965, Miyamoto and Matsumoto[72] and Miyamoto[73] demonstrated conclusively by serial thick (light microscopy) and thin (electron microscopy) sectioning that Negri bodies and nucleoprotein masses are identical. As in cell culture, these masses in which nucleoprotein is permanently trapped need not have precise margins and can contain entrapped host cell organelles and virus particles (Figure 13). Early inclusions contain nucleoprotein strands in rather dispersed arrays; the individual strands are often visible in such cases (Figure 14). Increasing condensation results in a progressive obscuring of individual strands which eventually results in the common granular appearance of inclusions (Figures 15 to 17). Further variations in the appearance of inclusions occur: these include 1) linear arrays of fine strands within the granular matrix, 2) bizarre tubular forms of varying diameter within the granular matrix, 3) convoluted membrane profiles intermixed with the granular matrix, and 4) multiple dense bodies within or at the margin of the granular matrix. The nature of these structures is not known. The spatial relationship between virus particle budding and inclusion body formation is not constant; with street rabies viruses and other lyssaviruses there is usually a close association (Figures 13 and 16), but with fixed rabies virus strains many inclusions are found without virus particle production upon nearby membranes.[16,74]

The course of rhabdovirus infections of cells in vivo varies greatly in speed and outcome. Some viruses destroy cells as quickly in vivo as in cell cultures; for example, intracerebrally inoculated vesicular stomatitis virus can kill suckling mice in 18 hr, at which time electron microscopic examination shows complete cytolysis of the brain. On the other hand, street rabies virus infection may take up to 20 to 30 days to kill mice or hamsters, and in terminal brain specimens most infected neurons have normal organelle structure (deformed only by the presence of inclusions bodies).[69,75] In those cases where extraneural organs are major targets of infection (e.g., vesicular stomatitis virus peripherally inoculated into mice or hamsters), ultrastructural cytopathology is usually severe.[71] In Table 1, under the heading "Disease in nature (experimental animals)," the term "encephalitis" is listed most commonly. This denotes that, even with many of the poorly characterized rhabdoviruses, there is evidence of the extremely destructive and inflammatory nature of central nervous system infection. In most cases, light and electron microscopic evidence contributed to this denotation, but further study of the nature of rhabdovirus infections is needed in every instance.

IV. AN ULTRASTRUCTURAL COMPARISON OF THE ANIMAL RHABDOVIRUSES

The thin-section electron micrographs presented in this chapter were selected as being representative of the major ultrastructural features of each of the animal rhabdoviruses. In this section these features will be summarized by reference to the micrographs and to the comparative data contained in Table 1. The viruses have been set in the same arbitrary morphologic categories used in the preceding sections.

A. *Vesiculovirus* Genus

The member viruses of the *Vesiculovirus* genus, vesicular stomatitis-Indiana, vesicular stomatitis-New Jersey, Cocal, Alagoas, Piry, Chandipura, and Isfahan viruses, are morphologically and morphogenetically indistinguishable.[14,29,49,76-78] The major ultrastructural features of infections in animal tissues and in cell cultures are (1) an extremely rapid course of infection (seen as near synchrony of synthetic and destructive events in sequential electron microscopic observations), (2) a devastating cytopathic change in infected cells, 3) a density of virus particle budding upon plasma membranes that is unmatched by any other rhabdoviruses, (4) an accumulation in extracellular spaces of very large numbers of virus particles which have an extremely uniform size and shape (as described in preceding sections), and (5) only a moderate accumulation of nucleoprotein inclusion bodies, which at best are diffuse or poorly marginated (Figures 9 to 11).

Taken together with the precisely parallel-sided cylindrical form of extracellular particles (Figure 12), these characteristics would be nearly pathognomonic of the genus; that is, an unknown or new virus would likely be identifiable as a vesiculovirus just by its overall appearance in thin sections of infected cell cultures or suckling mouse brain. In fact, such identification was made with Piry and Chandipura viruses before serological and physicochemical data were available.[89] The member viruses of the *Vesiculovirus* genus cannot be differentiated by any ultrastructural means; immunoelectron microscopy has not been used for comparative purposes.

B. *Lyssavirus* Genus

The member viruses of the *Lyssavirus* genus — rabies, Mokola, Lagos bat, and Duvenhage viruses — are morphologically and morphogenetically very similar.[16,21,22,68,69,79,80] Variations in the ultrastructral features of infections relate primarily to the level of adaptation by laboratory passage (e.g., street vs. fixed rabies virus strains).[16,74] In animal tissues and in cell cultures, the major ultrastructural features of infections caused by isolates directly from natural sources or in low passage are (1) a rather protracted course of infection (seen as a progressive involvement of more and more cells by sequential electron microscopy), (2) an extreme delay or absence of cytopathic changes in infected cells, (3) a progressive accumulation of nucleoprotein inclusion material which results first in diffuse or delicate array (Figures 13 and 14), and later in large, dense inclusion bodies that are unmatched by any other rhabdoviruses (Figures 15 to 18), (4) a focal occurrence of virus particle budding, which may be difficult to find upon plasma membranes (Figure 19) and may be prominent upon other membrane systems (Figures 13 and 20), (5) a variable capacity for accumulation of virus particles in extracellular spaces (e.g., the massive accumulation of rabies virus in salivary gland secretory lumina (Figure 21) vs. the sparsity of rabies virus particles in brain interstitium).

The most valuable characteristics for distinguishing laboratory adapted lyssaviruses have been their (1) greater propensity for budding upon plasma membranes, (2) increased capacity for causing ultrastructural cytopathology in infected cells in vitro and in vivo (and increased inflammatory response to cytopathology in vivo), and (3) common separation of sites of virus particle budding from sites of nucleoprotein inclusion formation.[74] Taken together with the uniform virus particle morphology usually obtained in thin sections (as described in preceding sections), analysis of these characteristics would allow a new or unknown rhabdovirus to be identified as a lyssavirus.[81-83]

For example, Mokola and Lagos bat viruses were initially related to rabies virus through comprehensive studies in which electron microscopy played a key role.[79] Similarly, electron microscopy has been used for the presumptive identification of laboratory (vaccine) strains of rabies virus in circumstances where vaccine-associated disease had to be distinguished from natural disease.[92]

C. Viruses With Cone-Shaped Particles

Two viruses, kotonkan and Obodhiang viruses, have been particularly difficult to relate to the other rhabdoviruses; despite their having cone-shaped particles, they have a reproducible distant serologic relationship to the lyssaviruses, all of which have uniformly cylindrical particle forms.[37, 38] This serologic anomaly has not been explained; however, in the present context these two viruses are considered together with bovine ephemeral fever virus, which is serologically unrelated but has identical morphologic characteristics.[39-42] The ultrastructural features of these three viruses in suckling mouse brain and in cell cultures are (1) a moderately fast course of infection, (2) a pronounced necrotizing cytopathology, (3) a focal and modest accumulation of virus nucleoprotein inclusions in infected cells. Most characteristic is the variable shape of particles as they bud upon plasma membranes and lie free in extracellular space; variable degrees of conification occur in all instances, but very broadly based particles are more common late in infection (Figures 22 to 27). Because particles have conical form while still budding, and have conical nucleocapsids (as described in preceding sections), it is unlikely that this shape variation is a simple preparation artifact. In any case, the described particle shape variation is so characteristic of these viruses that identification would be greatly helped by thin-section or negative contrast electron microscopy.

One additional cone-shaped virus, Marco virus, has been recognized recently (Figure 28).[91] The virus was originally isolated from a tropical lizard but grows in suckling mouse brain and mammalian cell cultures. In either substrate, the virus buds both upon plasma and intracytoplasmic membranes.

D. Viruses With Short Particle Length

The several rhabdoviruses with mean particle lengths significantly shorter than the prototype vesicular stomatitis virus do not constitute a homogeneous grouping (Figures 29 to 33). Their mean lengths vary between 130 and 167 nm, and the character of their infections also varies significantly. Kern Canyon (Figure 29), Barur (Figure 33), and Kununurra viruses have ultrastructural features which resemble those of vesicular stomatitis viruses; they have (1) a moderately fast infection course, (2) a necrotizing cytopathology, (3) a high density of virus particle budding upon plasma membranes of infected cells (Figure 29), (4) an extreme uniformity of virus particle size and shape, and (5) a minimal propensity for inclusion body formation.[35] Klamath virus particles have similar morphology, but the infection more closely resembles that caused by rabies virus.[84] Cytopathology in infected mouse brain is subtle, and virus buds upon intracytoplasmic membranes in association with prominent inclusion bodies. (Figures 30 to 32). In general, short particle length (in the absence of T-particle interference) and other ultrastructural characteristics might serve to narrow down the assignment of a new or unknown virus to a certain group, but serologic and physicochemical approaches would be required for overall characterization.

E. Viruses With Long Particle Length

The large number of animal rhabdoviruses which have particle lengths longer than that of vesicular stomatitis virus are brought together on a most abitrary basis. The viruses, Flanders, Hart Park, Mt. Elgon bat, Joinjakaka, Navarro, Kwatta and BeAn

157575, Mossuril and Kamese, Sawgrass and New Minto, and Chaco viruses (Figures 34 to 39) have been isolated from diverse sources in nature (Table 1). In some cases, they are represented by single isolates; because of their obscurity, lack of asociation with disease in man or domestic animals, and their poor growth in cell cultures, few studies have been undertaken.[36,85-88] The major ultrastructural features of infections in suckling mouse brain are (1) rather slow infection courses, (2) focal necrotizing cytopathology often centered upon the site of intracerebral inoculation, (3) a focalization of virus particle accumulation so that in a search for virus many areas of brain have barely detectable numbers of particles, and few sites have large groups of particles (Figures 34 and 35), and (4) often a significant accumulation of nucleoprotein inclusion material, which in the case of Mt. Elgon bat and Joinjakaka viruses resembles rabies inclusions[36] (Figure 36). There is no clear pattern of preference for budding from particular membrane systems among the long viruses (Fgures 37 to 39). In general, as with the short viruses, the finding of long particles by electron microscopy might serve to narrow down the placement of a new or unknown virus, but identification ultimately requires serologic and physicochemical study.

F. Uncharacterized Rhabdoviruses

New rhabdoviruses are being recognized with increasing frequency as electron microscopy is used more and more as part of comprehensive characterization schemes. Several new viruses have emerged from the category of "ungrouped arboviruses", and others from diverse searches for viruses in unusual settings. Rhabdovirus characteristics of some of these viruses have been recognized without reference to morphologic details. These newly recognized viruses include S-1643, Grey Lodge, Oita bat, Kimberley, Parry Creek, Yata, and Keuraliba viruses. Morphologic/morphogenetic characterization of each of these viruses is now being undertaken.

The fish and invertebrate rhabdoviruses are included in Table 1, but expansion on the listed morphologic characteristics of these viruses is beyond the scope of this chapter.

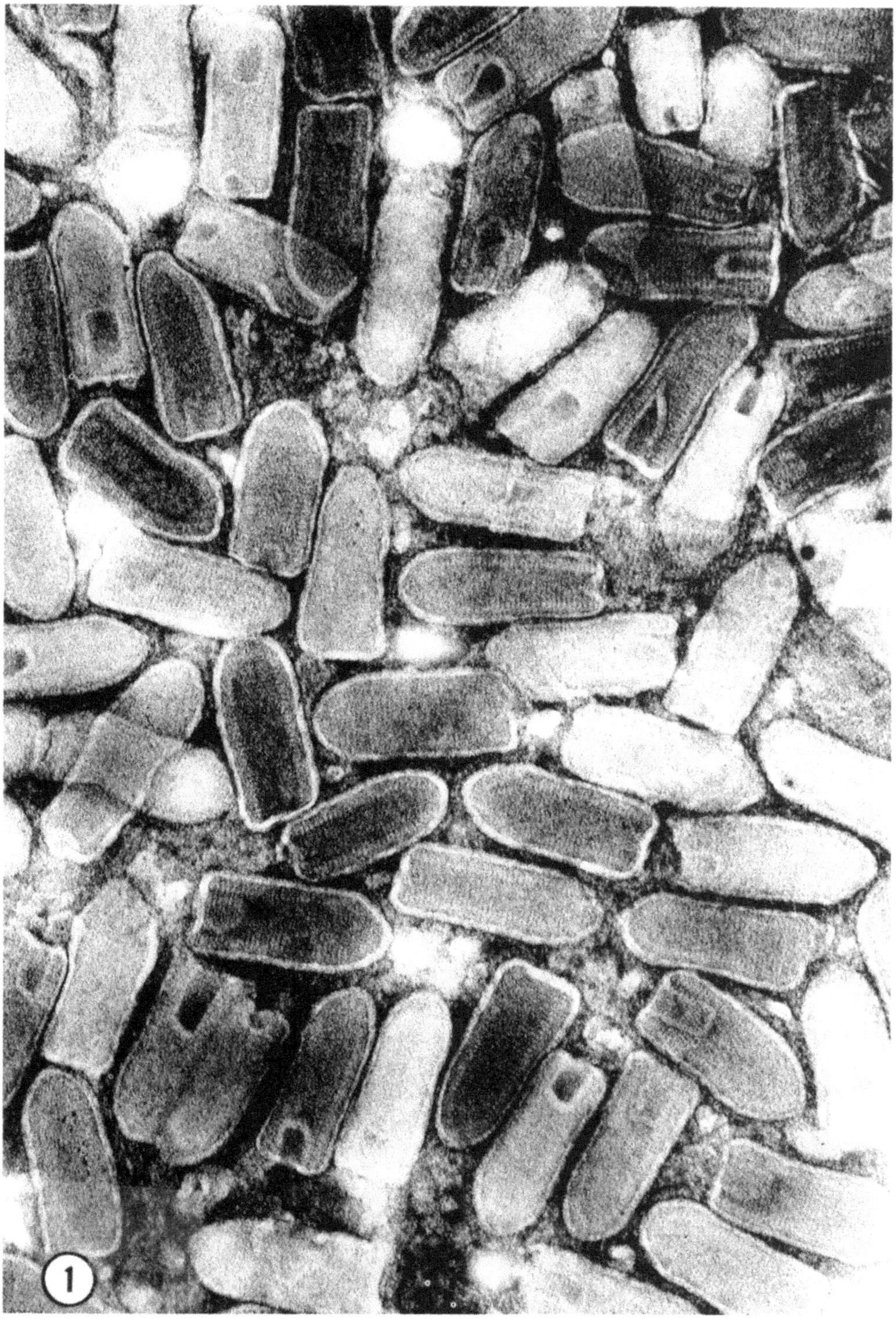

FIGURE 1. Purified preparation of vesicular stomatitis virus. Particles display a precisely cylindrical form. The envelope layer appears as a thickened line; the helically coiled nucleocapsid is seen as cross striations having an interval of 4.5 nm. (Magnification × 154,000.)

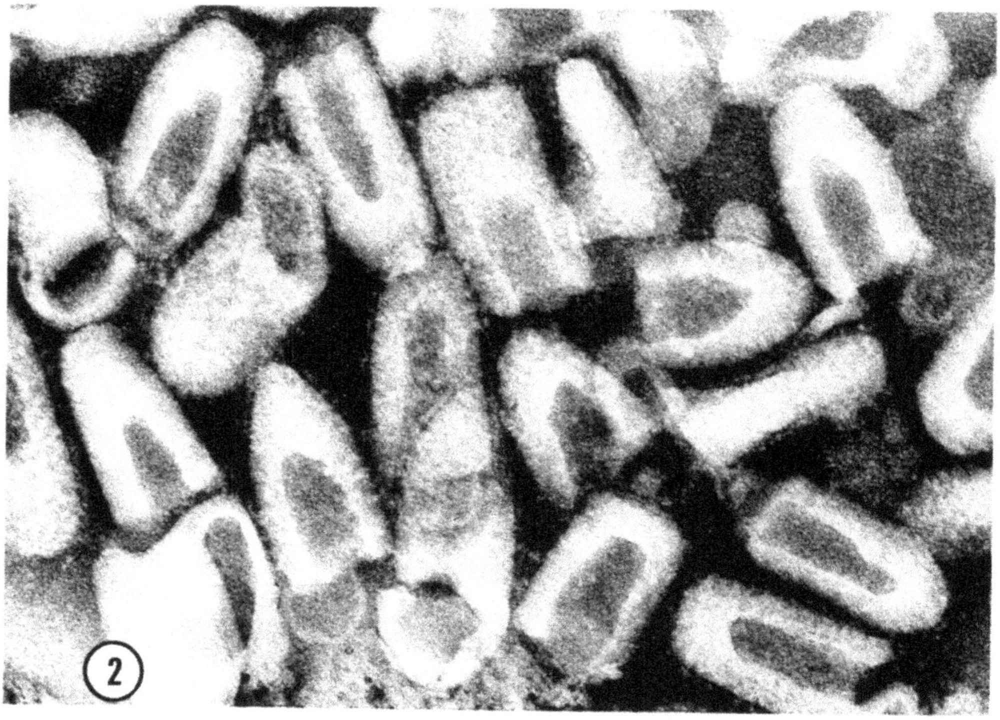

FIGURE 2. Rabies virus (Flury HEP strain). Some of the particles exhibit sixfold or honeycomb symmetry in the placement of the spikes; axial channels are especially prominent. (Magnification × 180,000.) Micrograph courtesy of Joan Crick, C. J. Smale, and F. Brown.

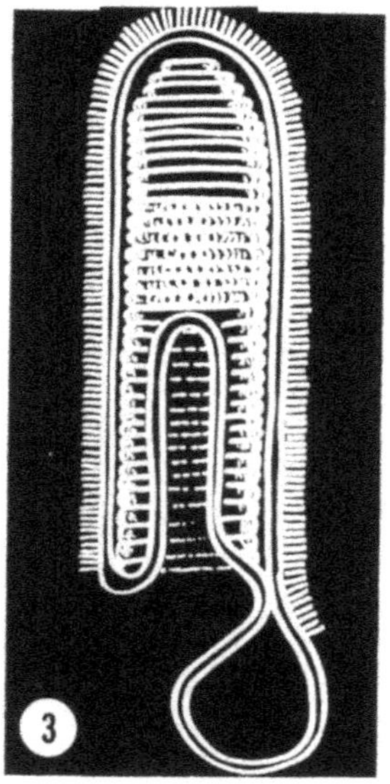

FIGURE 3. Model of an idealized rhabdovirus particle as seen with negative contrast electron microscopy, illustrating the surface projection layer, the unit membrane envelope (double lines) with invagination forming the axial channel and extra membrane forming a terminal bleb, and the helical ribonucleoprotein capsid.

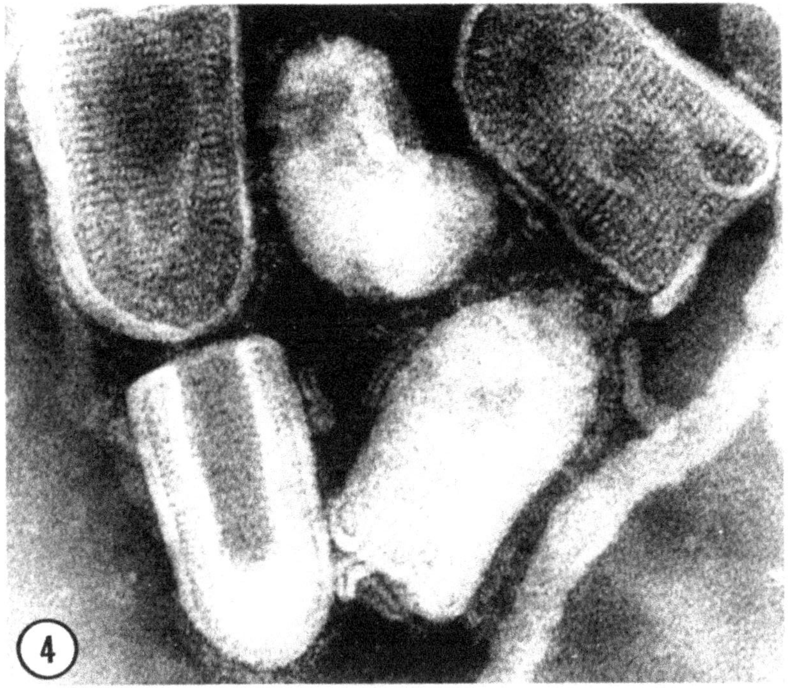

FIGURE 4. Kern Canyon virus. Surface projection layer, unit-membrane layer, and helically wound nucleocapsid are all evident in these particles which differ from the prototype *vesicular stomatitis virus only in their shorter length.* (*Magnification* × 302,500.)

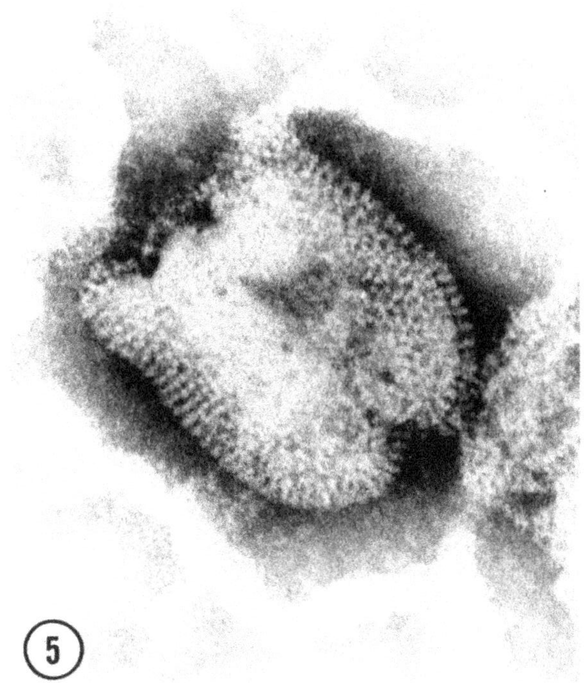

FIGURE 5. Klamath virus. Particles, lying side by side, exhibit a sixfold or honeycomb symmetry in the placement of spikes. (*Magnification* × 218,800.)

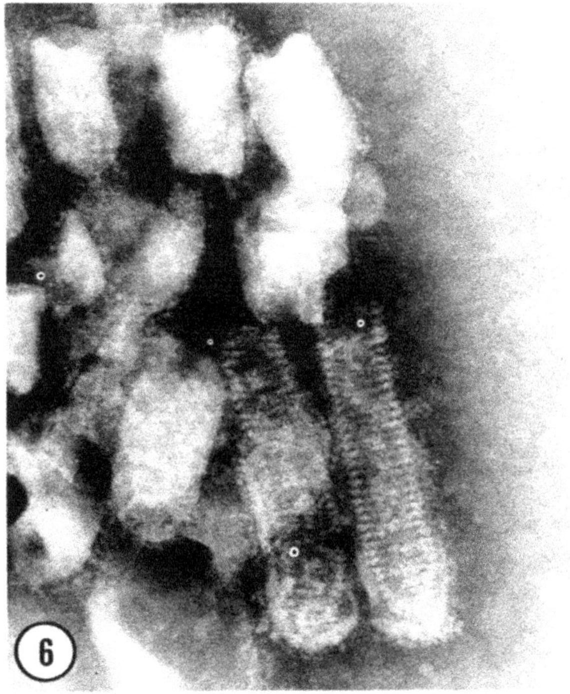

FIGURE 6. Mount Elgon bat virus is representative of viruses with cylindrical nucleocapsids which are longer than vesiculoviruses. (Magnification × 195,000.)

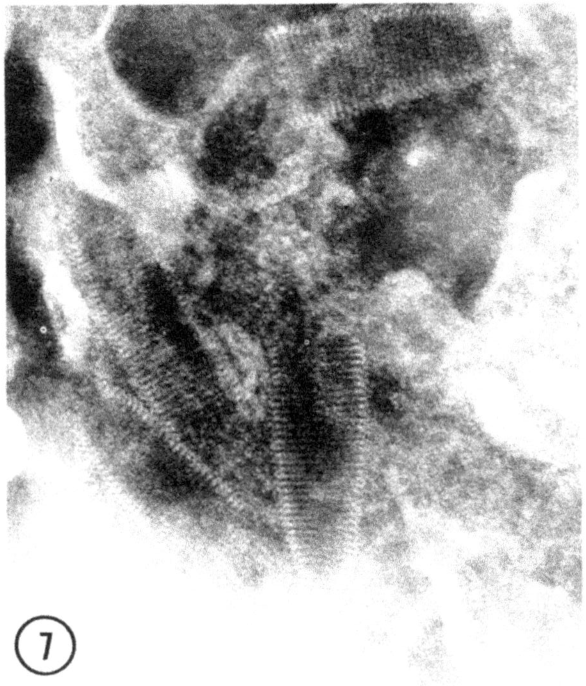

FIGURE 7. Kotonkan virus exhibits a slight broadening of the planar end. (Magnification × 178,000.)

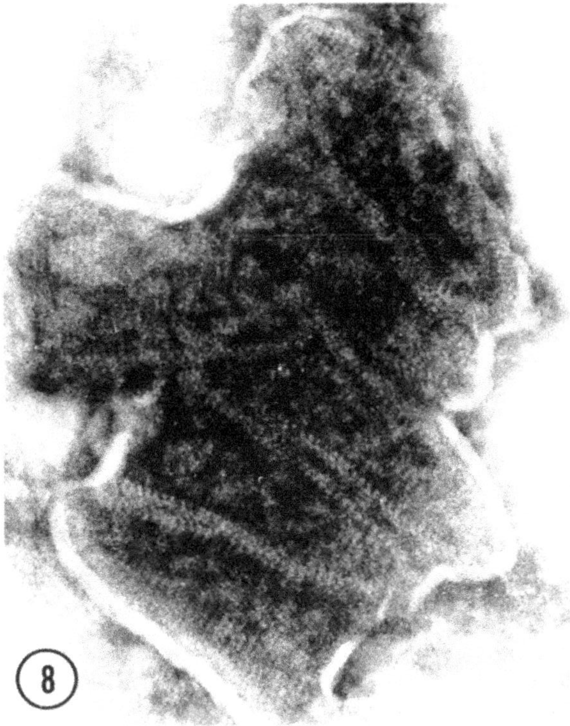

FIGURE 8. Bovine ephemeral fever virus. Virus particles display various degrees of broadening of the planar end. (Magnification × 270,000.)

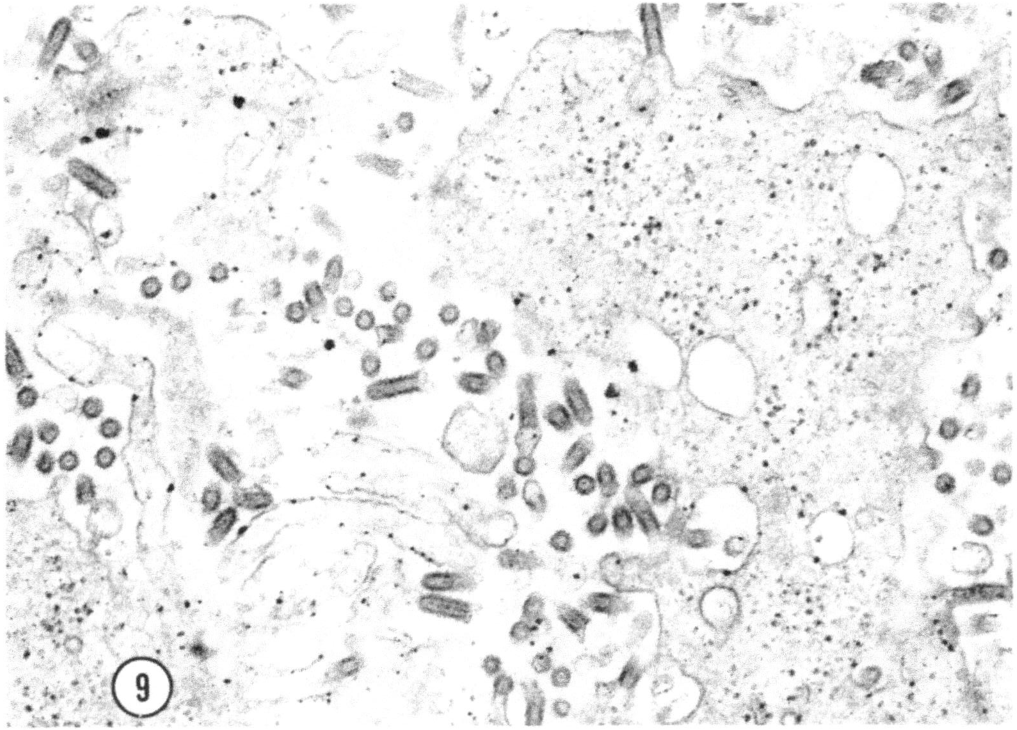

FIGURE 9. Vesicular stomatitis virus in mouse liver. Virus particles lie within intracytoplasmic vacuoles and bud upon the plasma membrane of an infected Kupffer cell, and lie free in the sinusoid. They exhibit characteristic bullet shape only when sectioned exactly longitudinally.(Magnification × 42,400.)

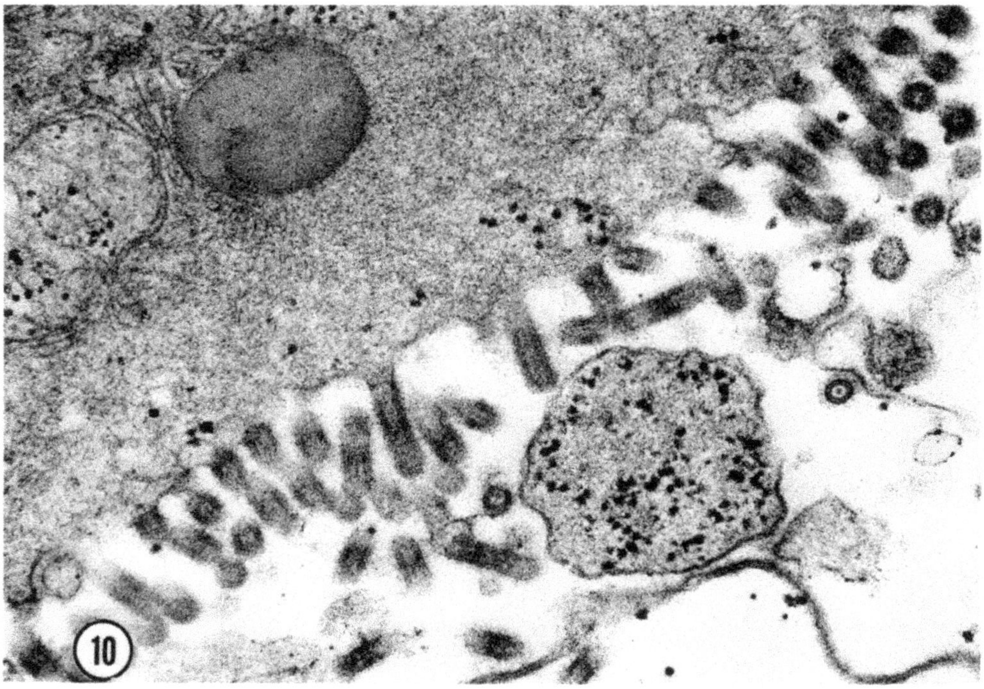

FIGURE 10. Chandipura virus in BHK-21 cell. Virus particles bud from the plasma membrane and lie free in extracellular space. They exhibit bullet shape in longitudinal section but appear elliptical or circular when sectioned obliquely or transversely. (Magnification × 49,100.)

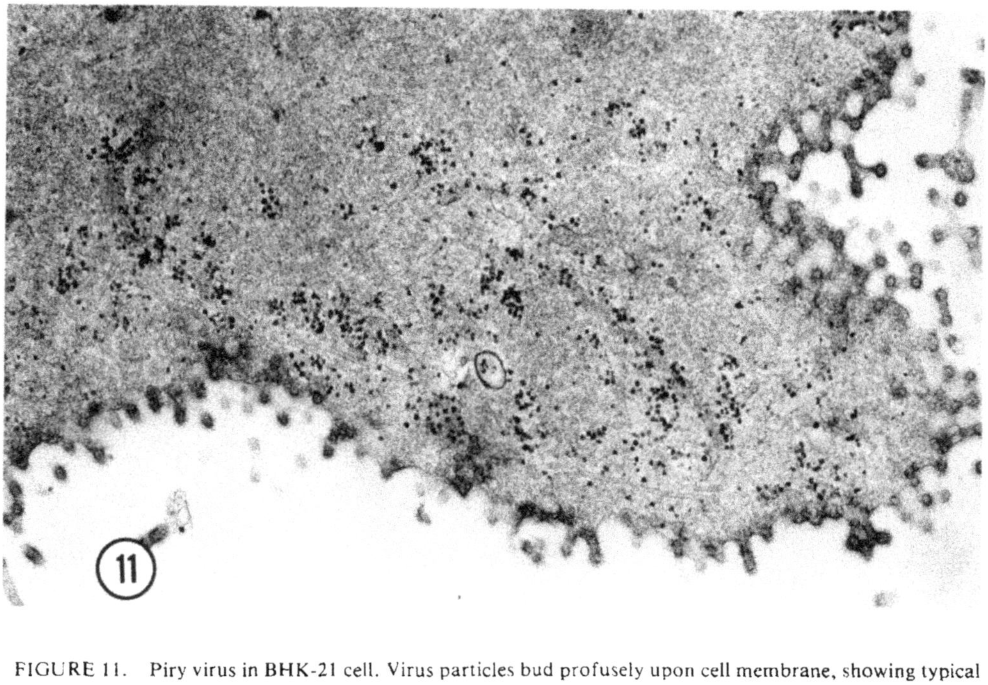

FIGURE 11. Piry virus in BHK-21 cell. Virus particles bud profusely upon cell membrane, showing typical structure depending on plane of section. (Magnification × 41,000.)

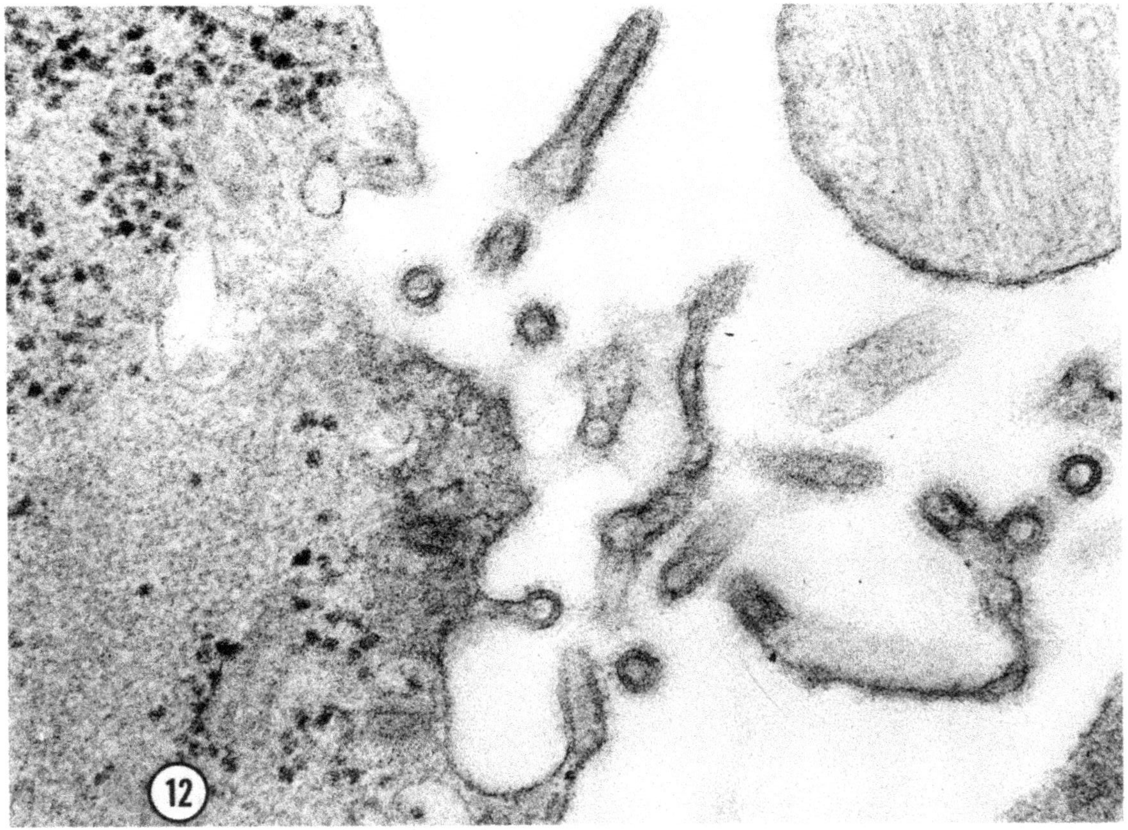

FIGURE 12. Higher magnification of Piry virus. The surface projection layer appears as a thick and delicate zone entirely surrounding virus particles; the nucleocapsid is seen as a dense beaded layer immediately beneath the envelope. (Magnification × 104,250.)

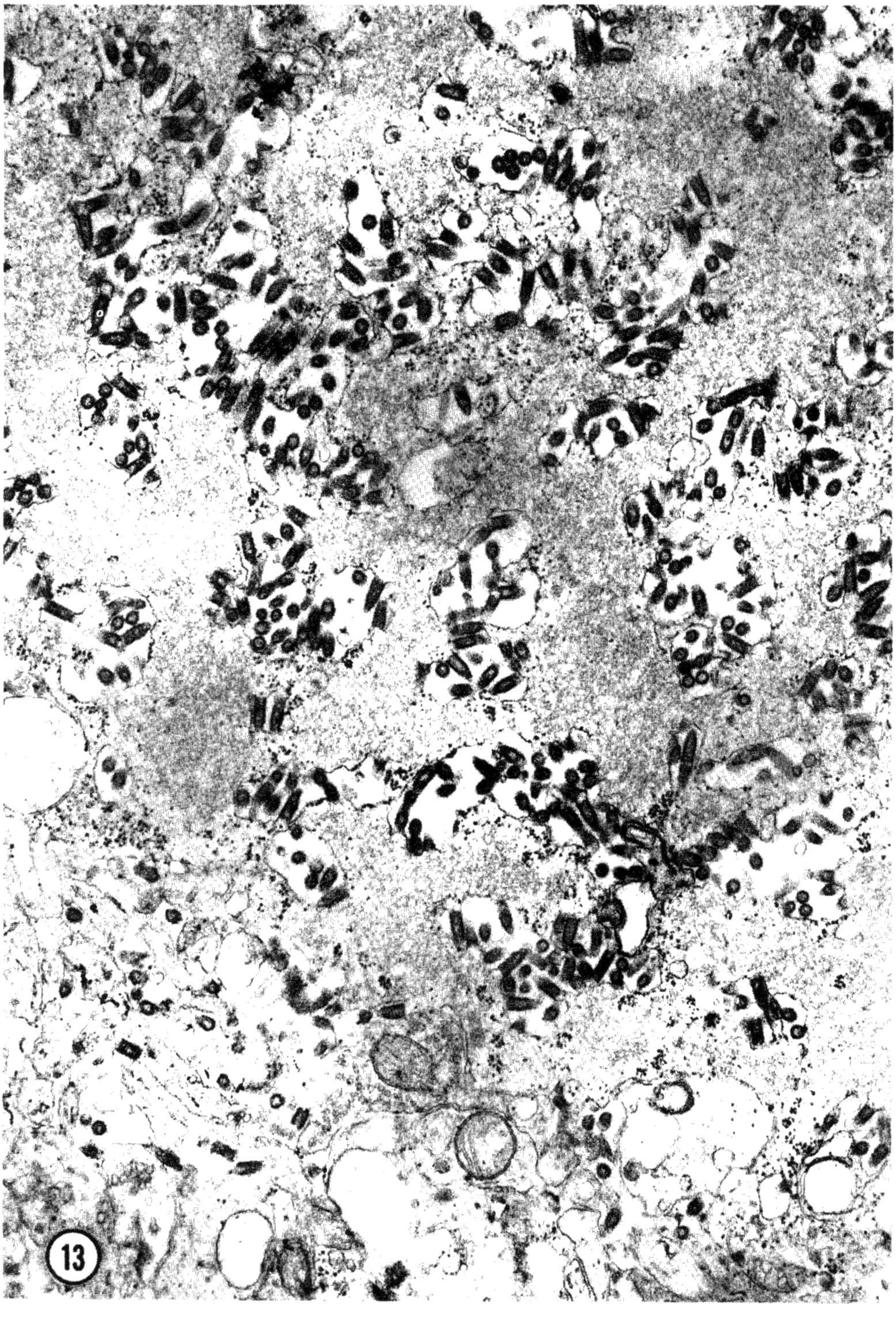

FIGURE 13. Mexican freetail bat rabies virus in hippocampal neuron of mouse. Large area of cytoplasm is filled with an inclusion body without precise margins and containing entrapped host cell organelles and numerous virus particles. (Magnification × 24,900.)

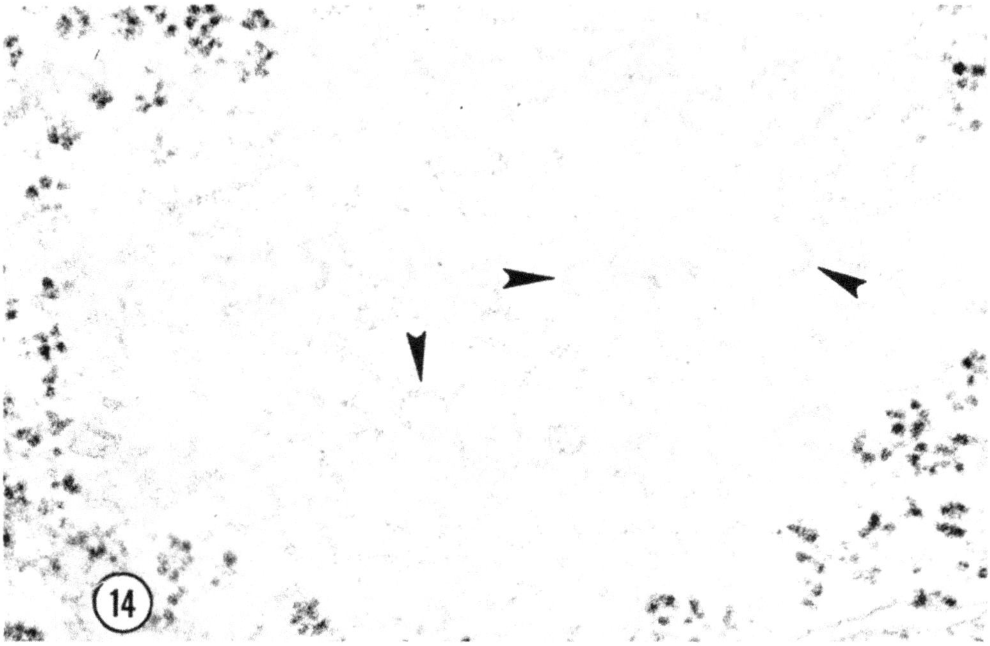

FIGURE 14. Lagos bat virus in neuron. An early inclusion body contains nucleoprotein strands in rather dispersed arrays; arrowheads point to substructural coiling of individual stands. (Magnification × 87,100.)

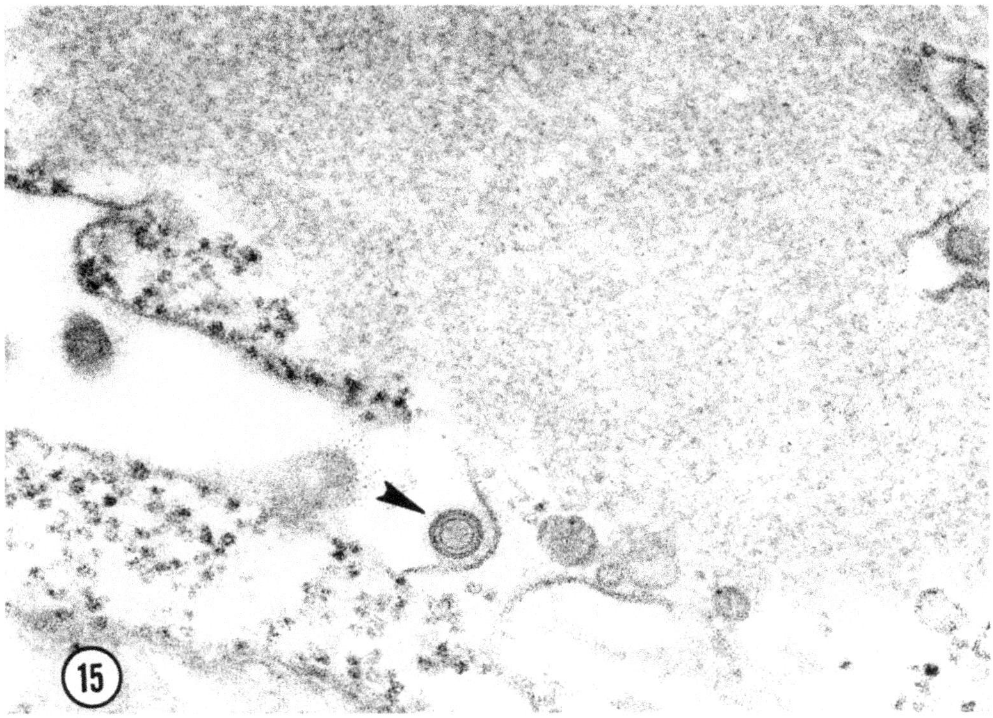

FIGURE 15. Rabies virus (street strain) in brain of skunk. Condensation has resulted in the common granular appearance of the inclusion body. The virus particle in cross section (arrowhead) shows the triple ring profile of the closely opposed nucleocapsid and envelope. (Magnification × 104,500.)

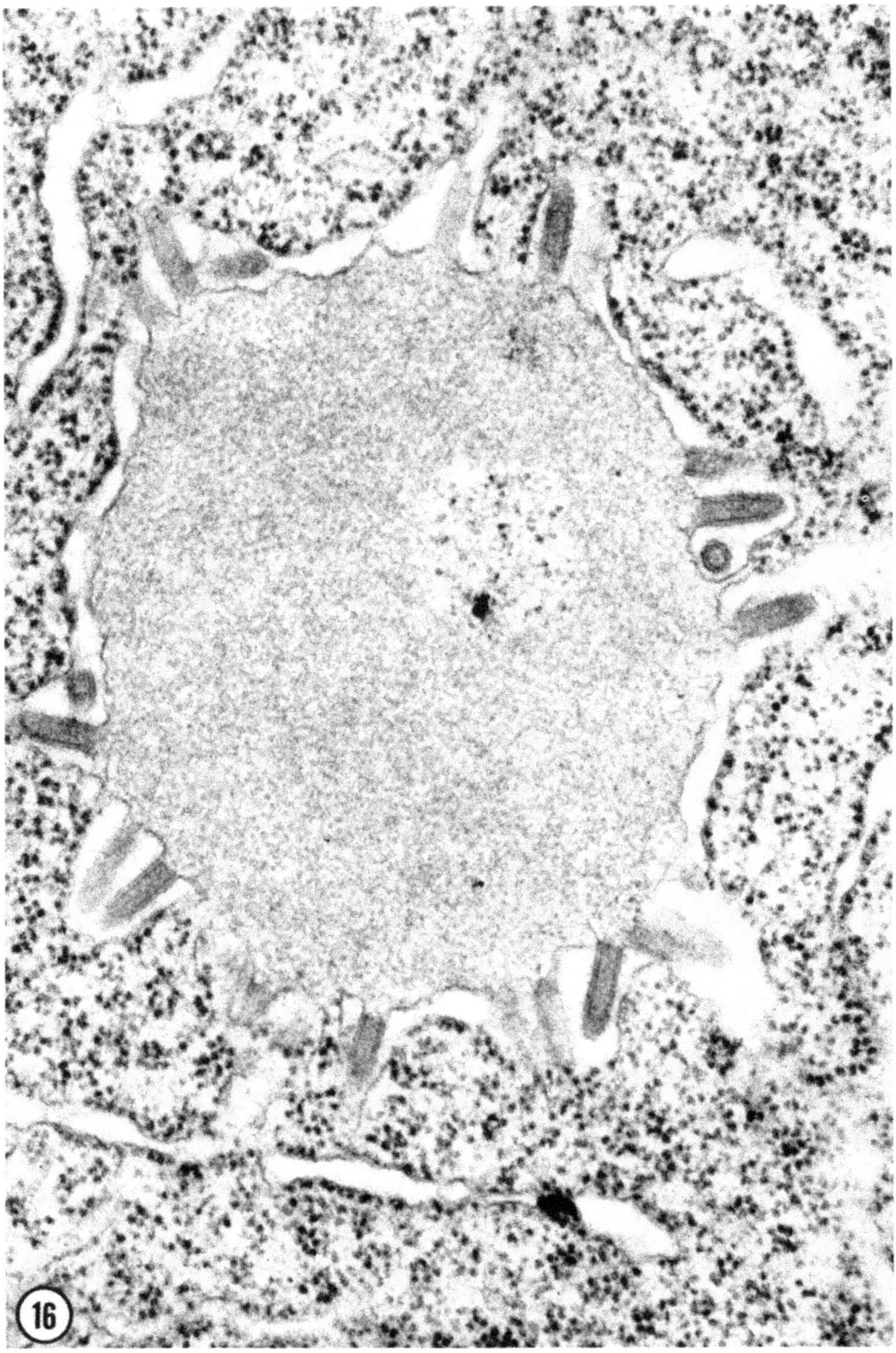

FIGURE 16. Lagos bat virus in a hippocampal neuron of a hamster. Virus particles bud upon intracytoplasmic membranes in close association with the densely granular appearing inclusion body. (Magnification × 64,000.)

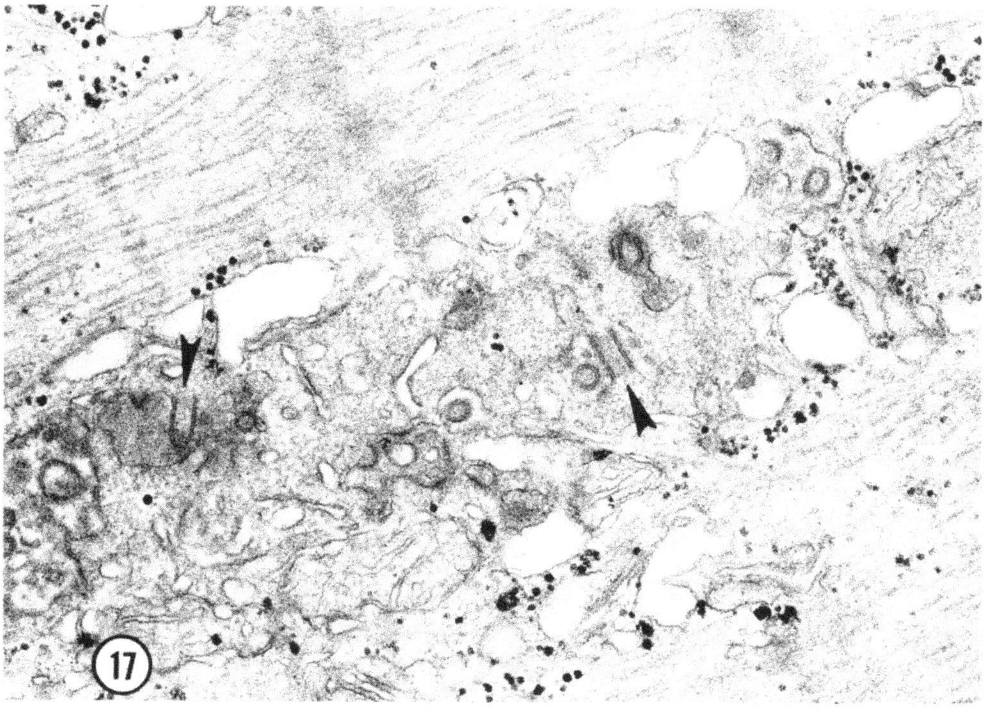

FIGURE 17. Mokola virus in masseter muscle of a hamster. There is a longitudinal displacement of the sarcoplasm by a dense inclusion body with associated virus particles. (Magnification × 47,500.)

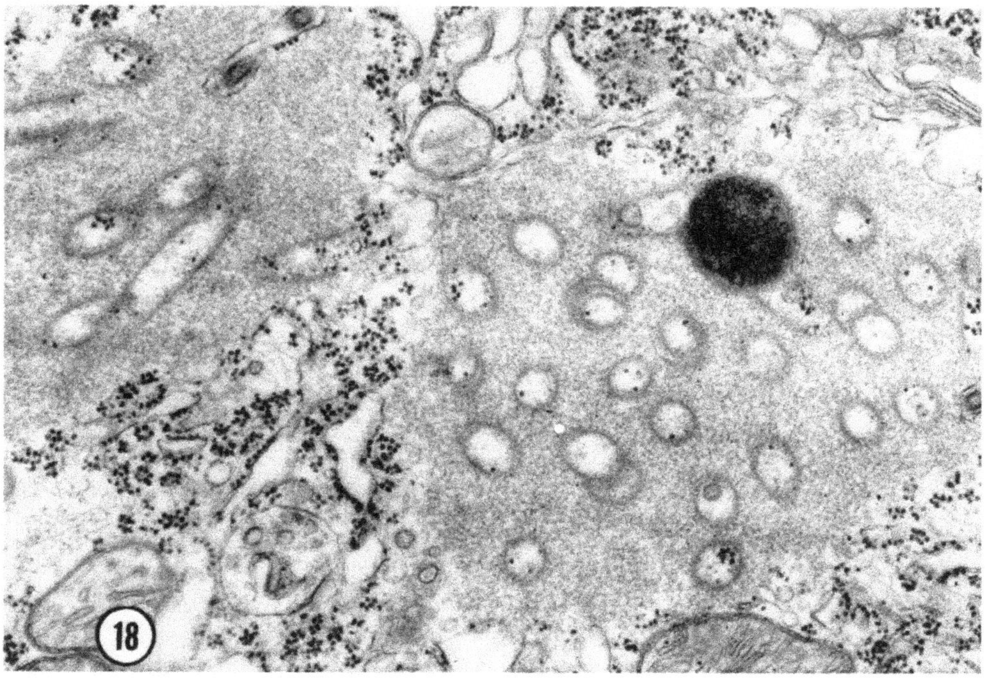

FIGURE 18. Rabies virus (*street virus; first passage from deer isolate*) in a thoracic spinal cord neuron of a mouse. Large, dense inclusion bodies with complex tubular structures *commonly found with street virus* strains in vivo fill the cytoplasm. (Magnification × 30,000.)

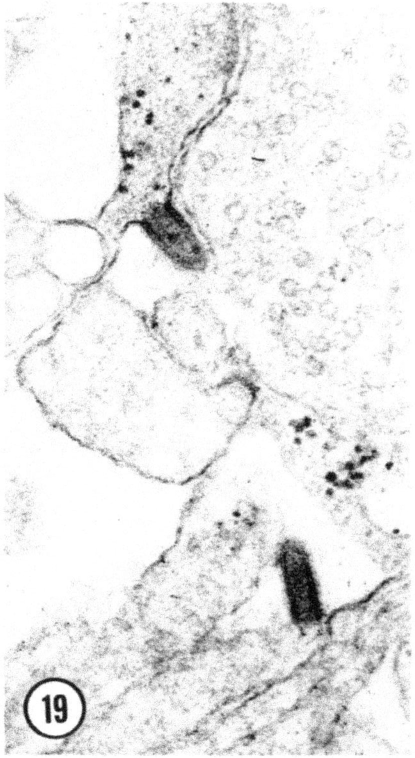

FIGURE 19. Duvenhage virus in mouse brain. A rare virus particle is seen budding upon a plasma membrane. (Magnification × 67,000.)

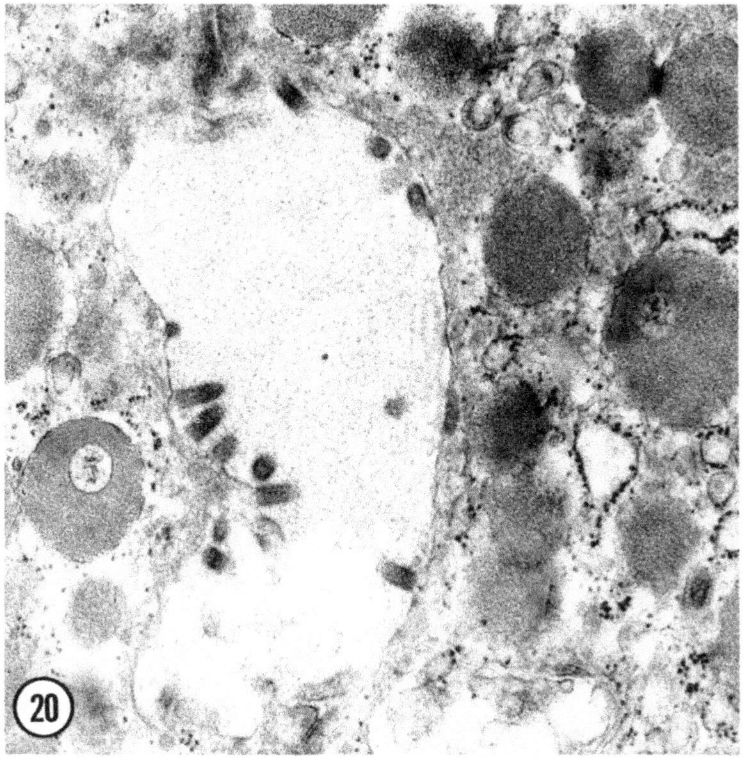

FIGURE 20. Rabies virus in fox salivary gland. Virus particles bud upon the plasma membrane of two mucogenic acinar cells into an intercellular canaliculus. (Magnification ×35,000.)

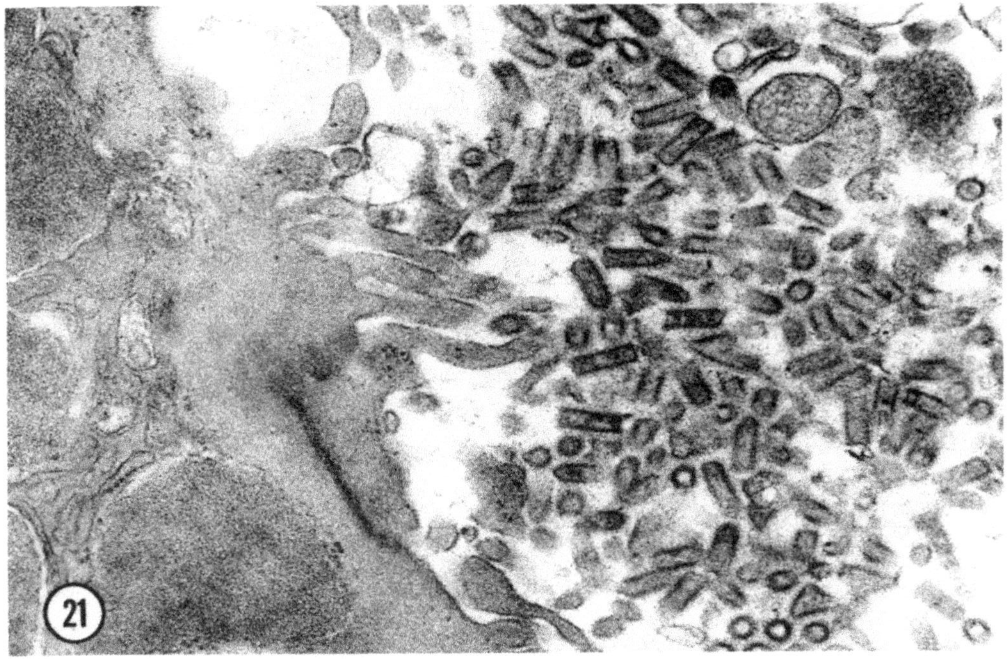

FIGURE 21. Rabies virus in fox salivary gland. There is a massive accumulation of virus particles in the salivary gland lumen. (Magnification × 46,500.)

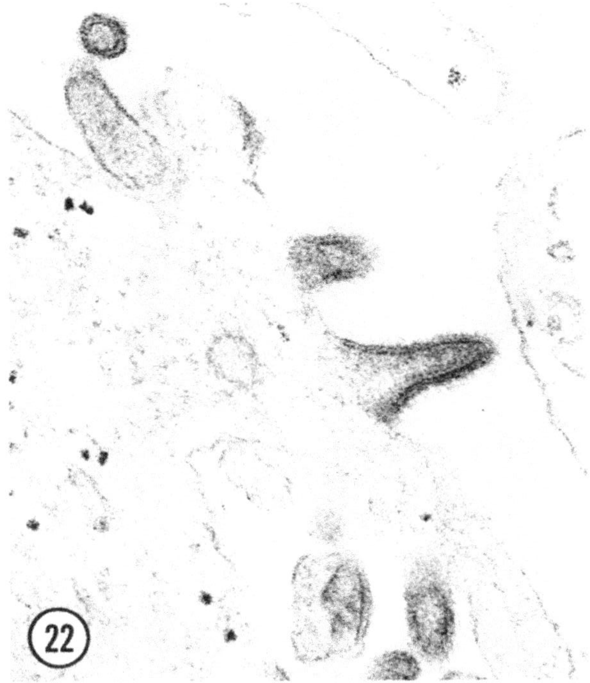

FIGURE 22. Kotonkan virus in mouse brain. Virus particles bud upon the plasma membrane of a neuronal process; one particle illustrates broadening of the base, presenting a conical form. (Magnification × 87,400.)

FIGURE 23. Kotonkan virus. Characteristic variability of shape of particles can be seen as they bud upon plasma membranes. (Magnification × 88,400.)

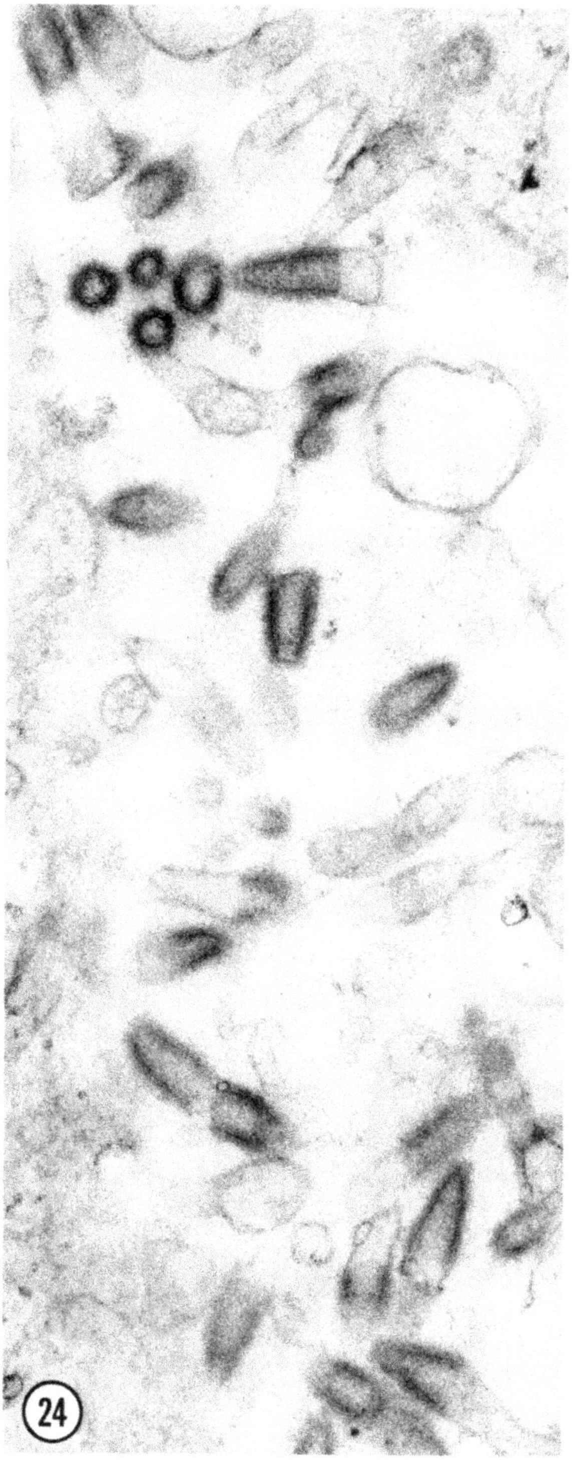

FIGURE 24. Obodhiang virus in mouse brain. Variability of shape of virus particles is apparent. (Magnification × 79,000.)

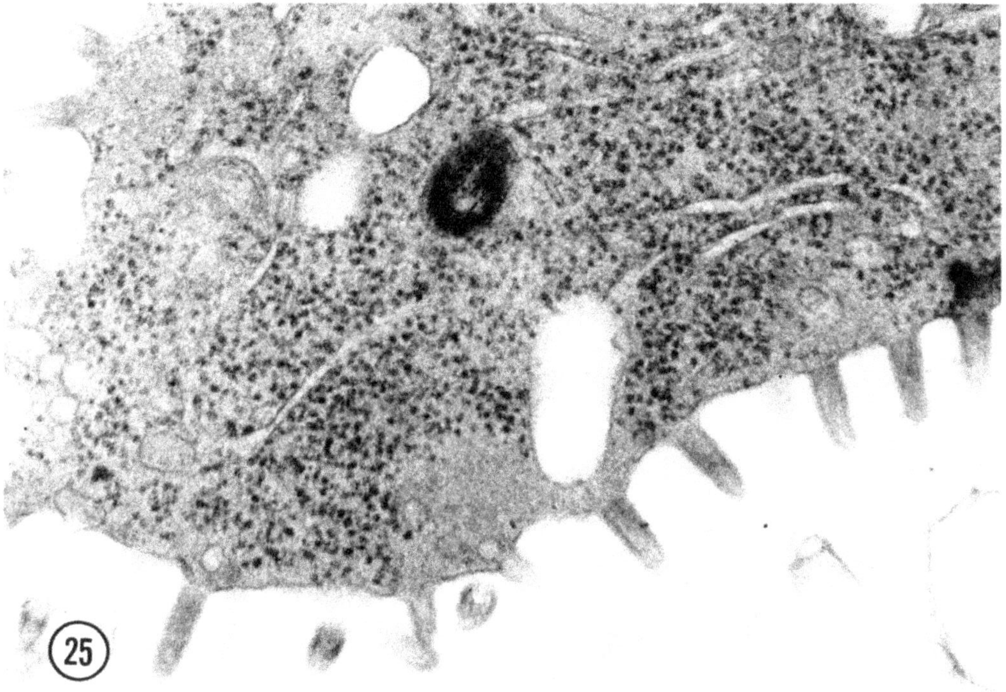

FIGURE 25. Bovine ephemeral fever virus in vero cell. Most particles are cone-shaped with narrow base and nearly parallel sides. (Magnification × 56,700.)

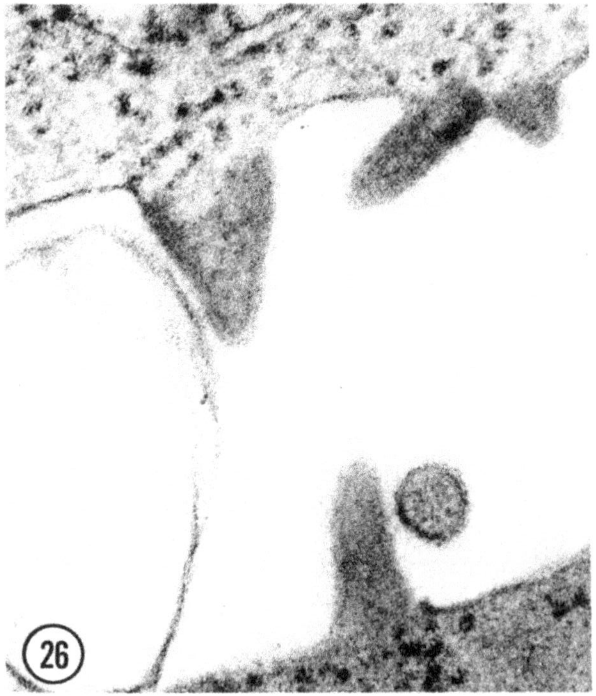

FIGURE 26. Bovine ephemeral fever virus in vero cells. Particles range in shape from a parallel-sided bullet to a broad-based cone. (Magnification × 110,000.)

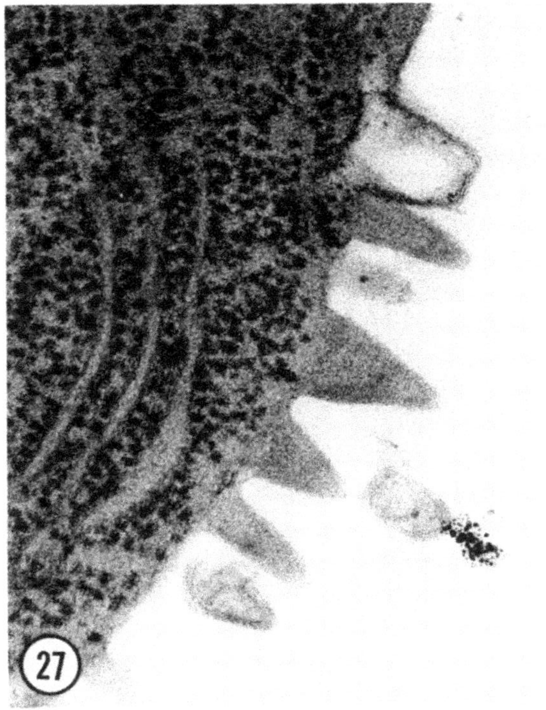

FIGURE 27. Bovine ephemeral fever virus in vero cell. Late in infection many of the virus particles have the broad-based cone shape and bud freely upon the plasma membrane. Magnification × 91,000.)

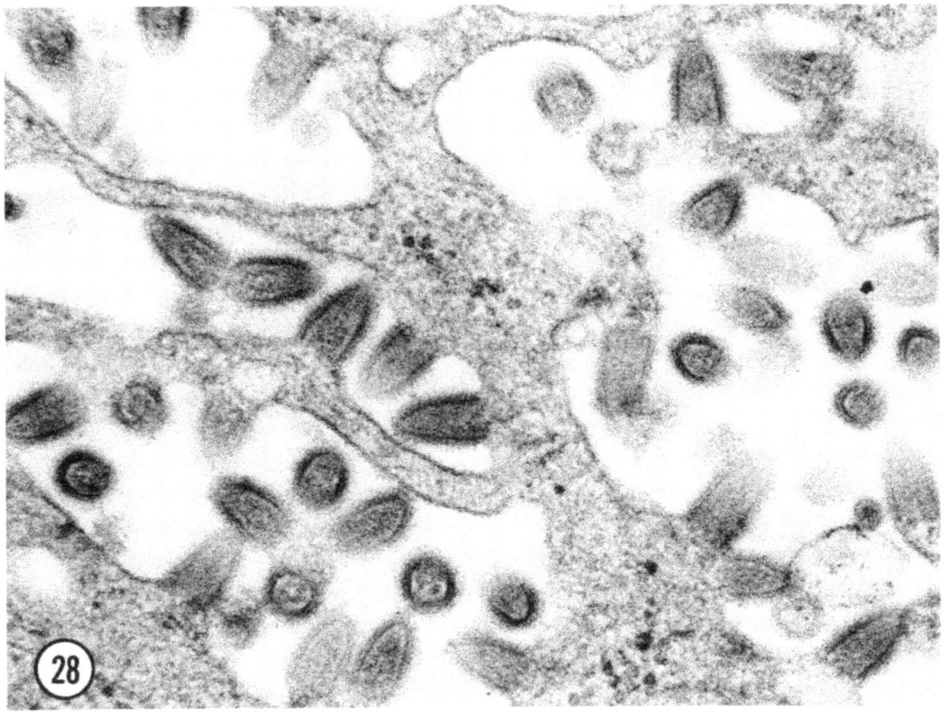

FIGURE 28. Marco virus in turtle heart (TH-1) cells. The cone-shaped virus buds freely upon both intracytoplasmic and plasma membranes. (Magnification × 84,400.)

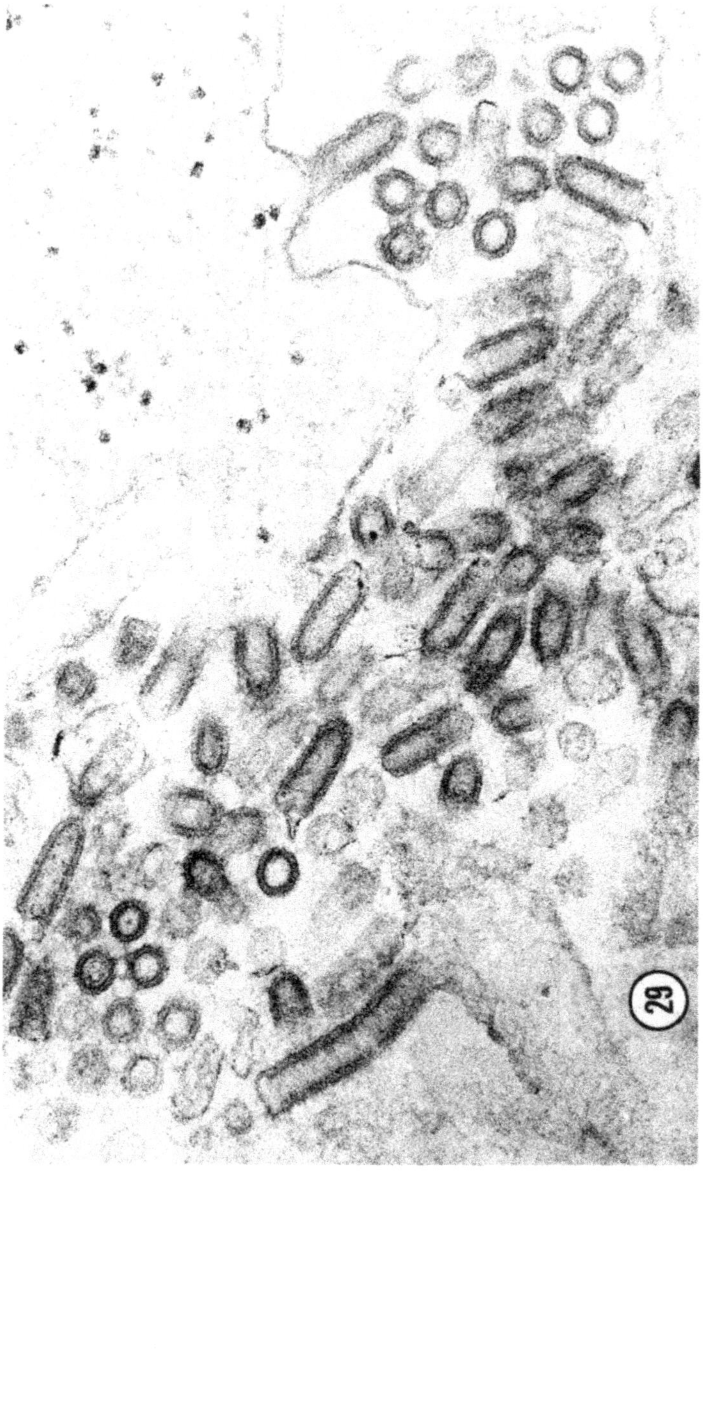

FIGURE 29. Kern Canyon virus in mouse brain. Virus particles bud upon plasma membranes of neuronal processes and accumulate in the intercellular spaces. They possess the uniform bullet shape, differing only from the prototype vesicular stomatitis virus in their shorter length. (Magnification × 84,400.)

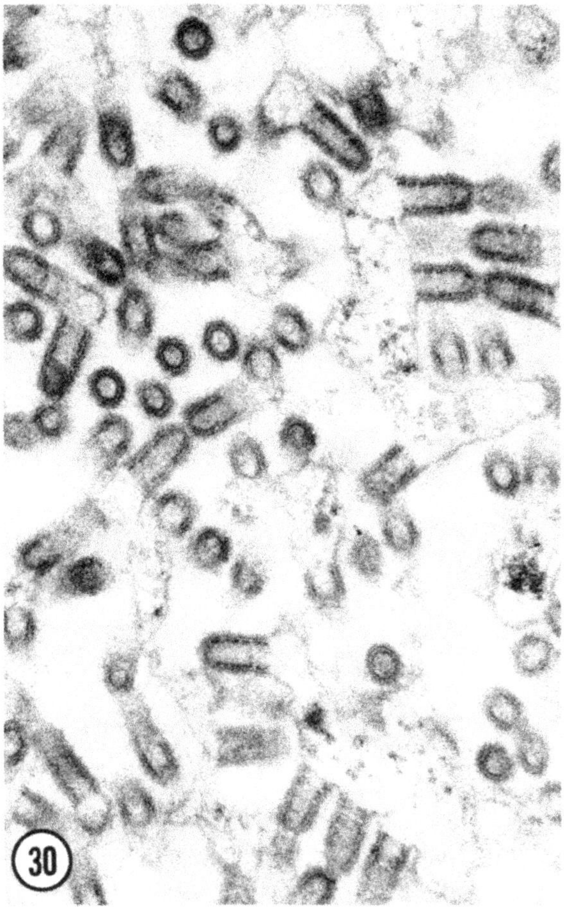

FIGURE 30. Klamath virus in BHK-21 cell. The uniformly short, parallel-sided virus particles bud upon intracytoplasmic membranes and accumulate within distended cisternae. (Magnification × 83,000.)

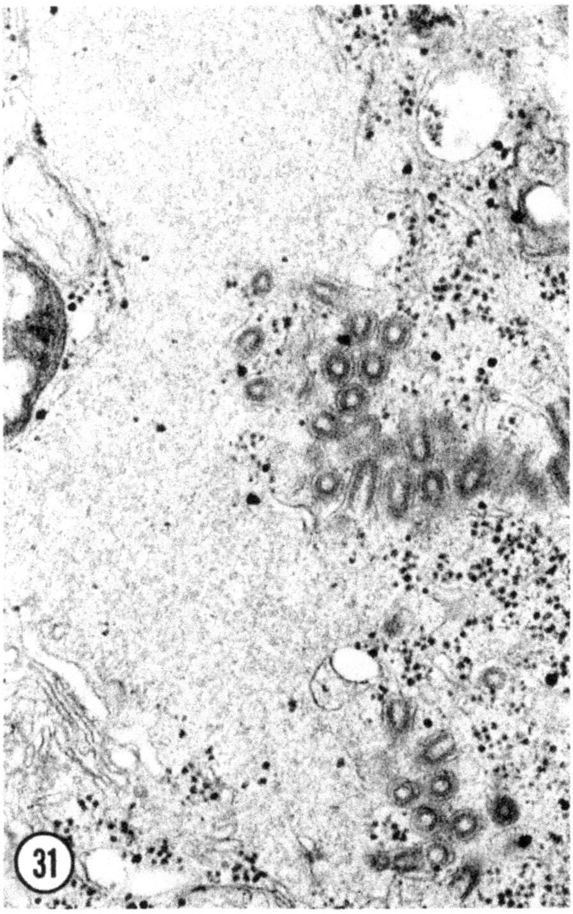

FIGURE 31. Klamath virus in mouse brain. It differs from the other short particles in that the virus buds upon intracytoplasmic membranes in association with prominent inclusion bodies. (Magnification × 49,200.)

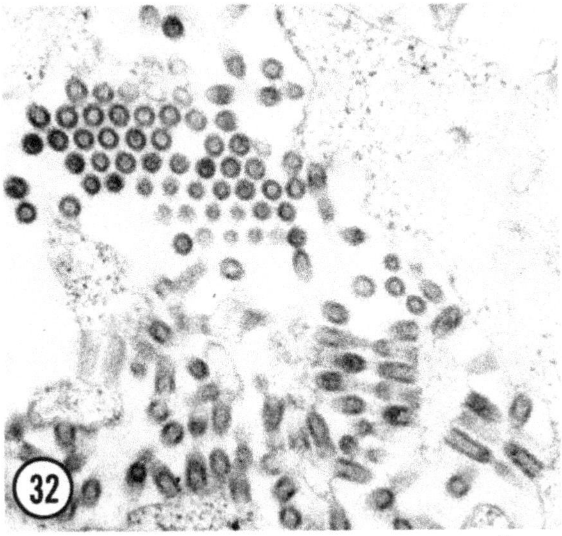

FIGURE 32. Klamath virus in BHK-21 cell. Virus particles are closely packed in paracrystalline array. (Magnification × 45,200.)

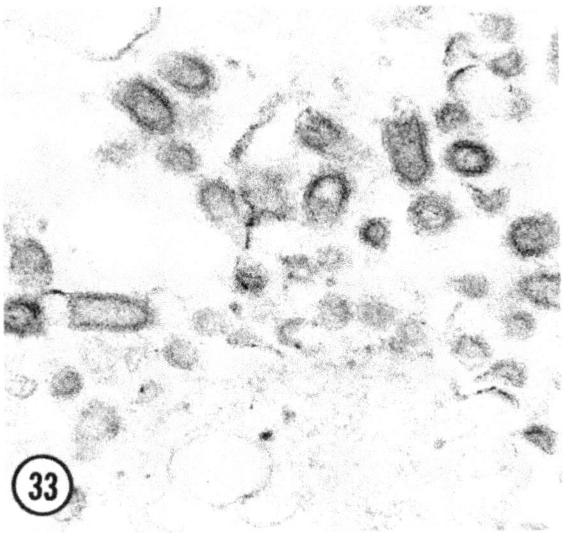

FIGURE 33. Barur virus in mouse brain. Characteristic short shaped virus particles accumulate in extracellular space. (Magnification × 84,400.)

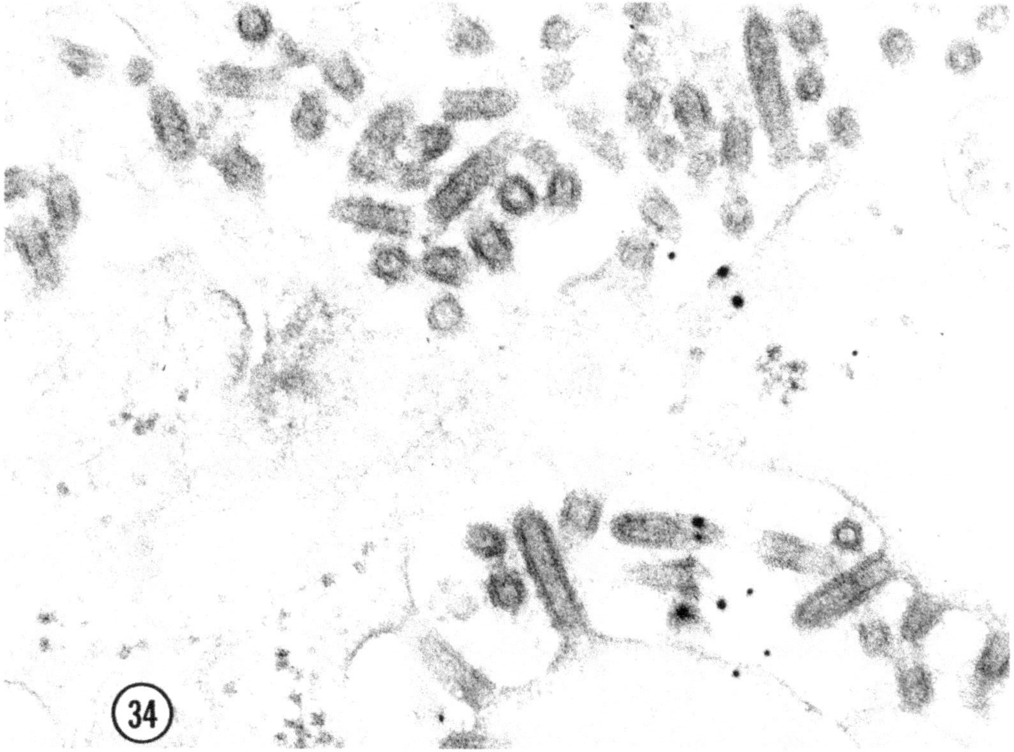

FIGURE 34. Flanders virus in mouse brain. Focal accumulation of long virus particles bud from plasma membrane and accumulate in extracellular spaces. (Magnification × 84,400.)

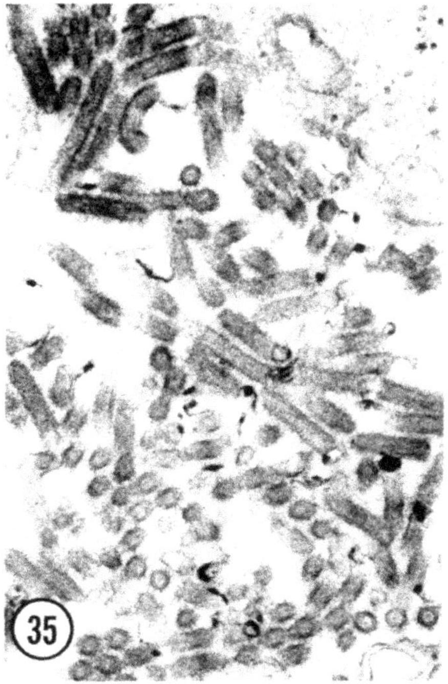

FIGURE 35. Mount Elgon bat virus in mouse brain. Virus particles in proper plane of section display cross striations and longer length than the prototype virus. (Magnification × 51,700.)

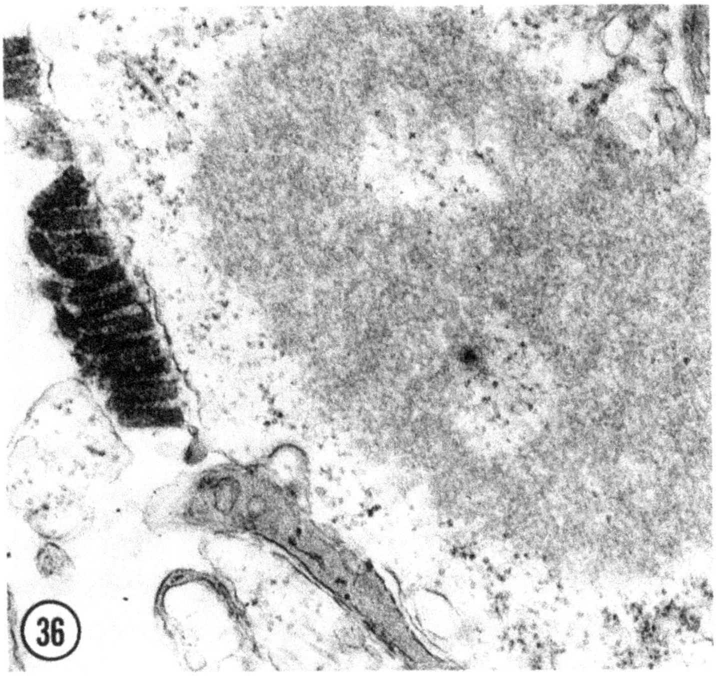

FIGURE 36. Joinjakaka virus in mouse brain. The long virus particles are lined up close to the plasma membrane. The inclusion body is similar to that found in rabies. (Magnification × 49,400.)

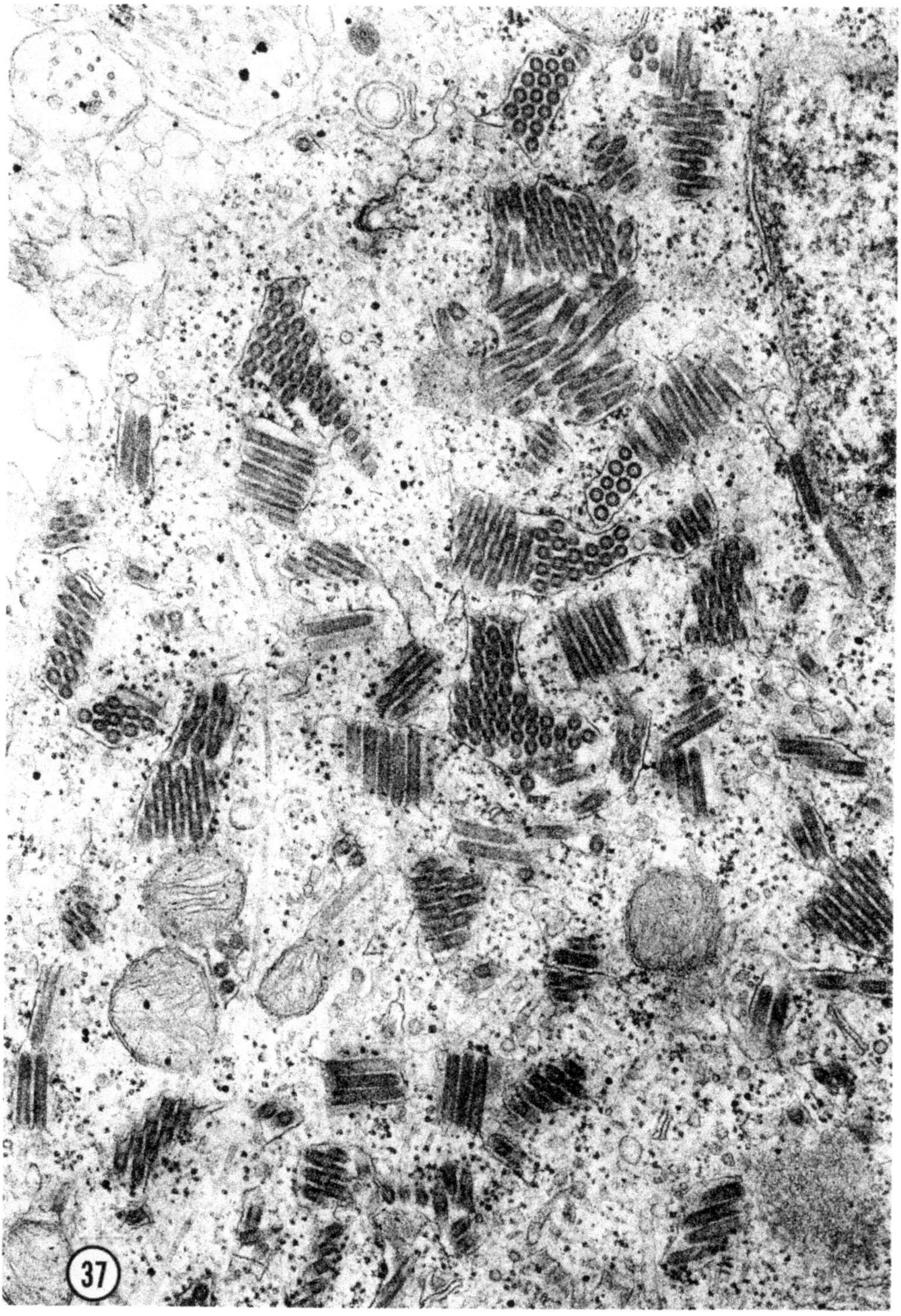

FIGURE 37. Navarro virus in mouse brain. Navarro virus shows a propensity for budding upon intracytoplasmic membranes and closely packs the distended vacuoles of an infected neuron. (Magnification × 40,250.)

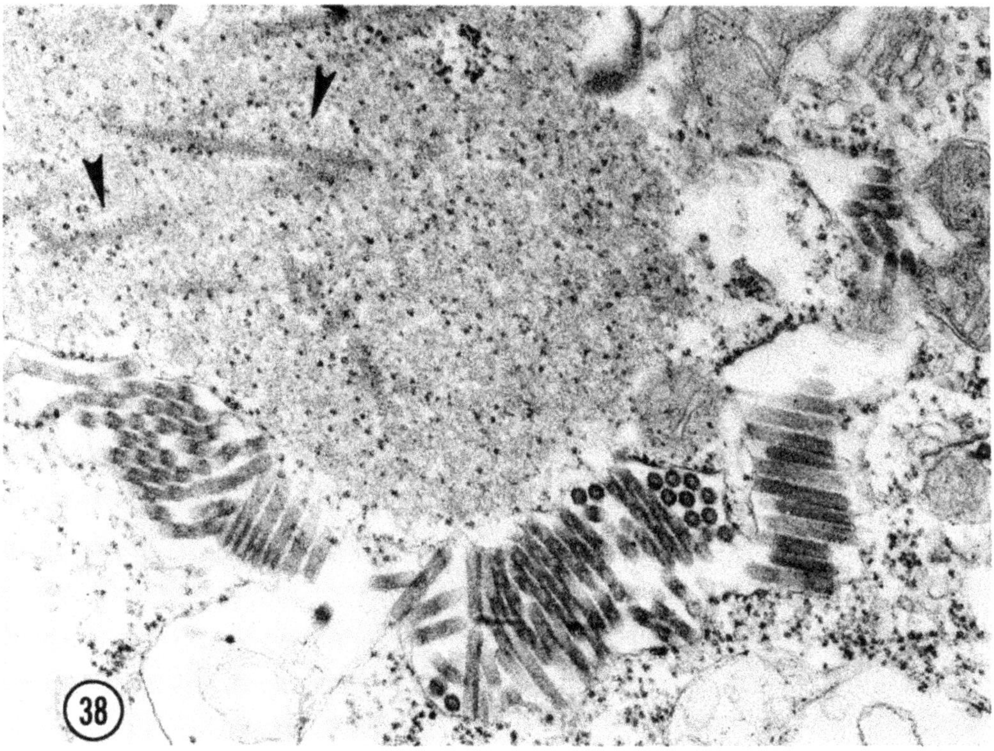

FIGURE 38. Kwatta virus in mouse brain. Accumulations of virus intracytoplasmic vacuoles lie in close proximity to large inclusion body containing long filaments (arrowheads). (Magnification × 43,500.)

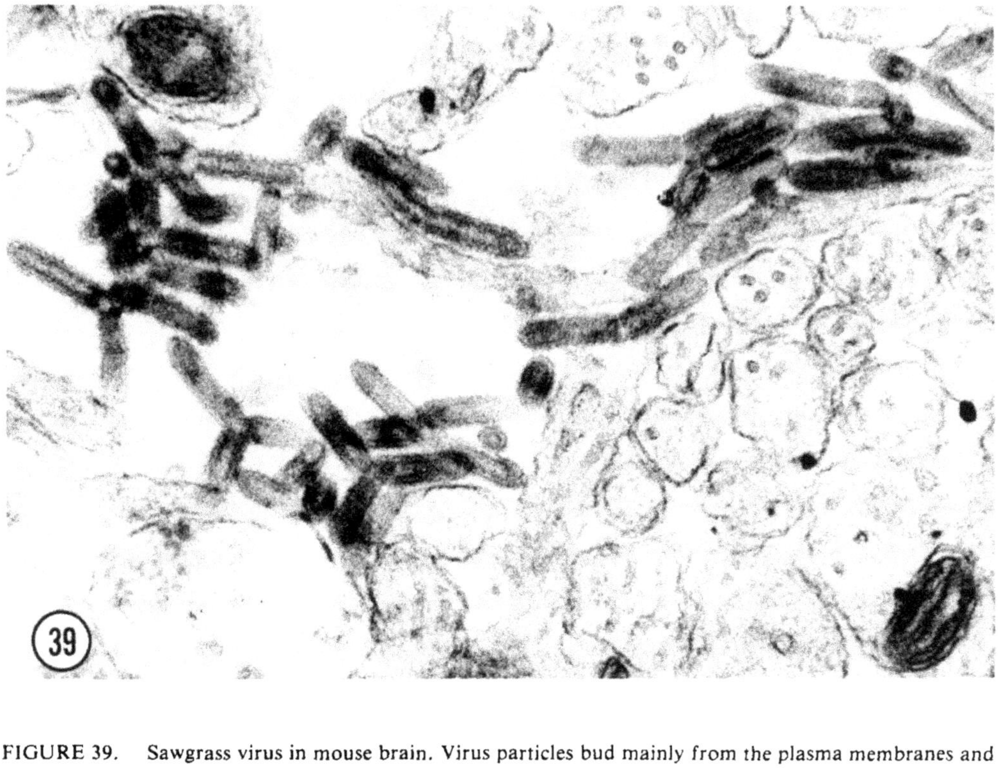

FIGURE 39. Sawgrass virus in mouse brain. Virus particles bud mainly from the plasma membranes and accumulate in extracellular spaces. (Magnification × 69,700.)

REFERENCES

1. **Melnick, J. L. and McCombs, R. M.**, Classification and nomenclature of animal viruses, *Prog. Med. Virol.*, 8, 400, 1966.
2. **Wildy, P.**, Classification and nomenclature of viruses. First report of the International Committee on Nomenclature of Viruses, *Monogr. Virol.*, 5, 1, 1971.
3. **Fenner, F.**, Classification and nomenclature of viruses. Second report of the International Committee on Taxonomy of Viruses, *Intervirology*, 7, 1, 1976.
4. **Chow, T. L., Chow, F. H., and Hanson, R. P.**, Morphology of vesicular stomatitis virus, *J. Bacteriol.*, 68, 724, 1954.
5. **Bradish, C. J., Brooksby, J. B., and Dillon, J. F.**, Biophysical studies of the virus system of vesicular stomatitis, *J. Gen. Microbiol.*, 14, 290, 1956.
6. **Reczko, E.**, Elektron Mikroskopische Untersuchungen am Virus des Stomatitis Vesicularis, *Arch. Gesamte Virusforsch.*, 10, 588, 1960.
7. **Howatson, A. F. and Whitmore, G. F.**, The development and structure of vesicular stomatitis virus, *Virology*, 16, 466, 1962.
8. **Stone, R. S., Sellers, M. I., and Hiramoto, T.**, Morphology of the development and release of vesicular stomatitis virus, *J. Appl. Physiol.*, 32, 1639, 1961.
9. **Almeida, J. D., Howatson, A. F., Pinteric, L., and Fenje, P.**, Electron microscope observations on rabies virus by negative staining, *Virology*, 18, 147, 1962.
10. **Pinteric, L., Fenje, P., and Almeida, J. D.**, The visualization of rabies virus in mouse brain, *Virology*, 20, 208, 1963.
11. **Davies, M. C., Englert, M. E., Sharpless, G. R., and Cabasso, V. J.**, The electron microscopy of rabies virus in cultures of chicken embryo tissues, *Virology*, 21, 642, 1963.
12. **Atanasiu, P., Lépine, P., and Dragonas, P.**, Etude cinétique du virus rabique en culture de tissus à l' aide des anticorps fluorescents et des coupes ultrafine, *Ann. Inst. Pasteur, Paris*, 105, 813, 1963.
13. **Roots, E. and Schultze, I. M.**, Neuere elektronenmikroskopische Befunde an Gehiren nach Infektion mit Tollwutvirus, *Zentralbl. Bakteriol. Parasitenkd. Infektionskr. Hyg. Abt. 1: Orig.*, 188, 159, 1963.
14. **Howatson, A. F.**, Vesicular stomatitis and related viruses, *Adv. Virus Res.*, 16, 196, 1970.
15. **Knudson, D. L.**, Rhabdoviruses, *J. Gen. Virol.*, 20, 105, 1973.
16. **Murphy, F. A.**, Morphology and morphogenesis, in *The Natural History of Rabies*, Vol. 1, Baer, G. M., Ed., Academic Press, New York, 1975, chap. 2.
17. **Murphy, F. A.**, Evolution of rhabdovirus tropisms, in *Viruses, Evolution and Cancer*, Kurstak, E. and Maramorosch, K., Eds., Academic Press, New York, 1974, 699.
18. **Bishop, D. H. L. and Smith, M. S.**, Rhabdoviruses, in *The Molecular Biology of Animal Viruses*, Vol. 1, Nayak, D. P., Ed., Marcel Dekker, New York, 1977, chap. 5.
19. **Berge, T. O., Ed.** *International Catalogue of Arboviruses*, 2nd ed., Public Health Service, U.S. Department of Health, Education, and Welfare, Washington, D.C., 1975, 9.
20. **Wagner, R. R.**, Reproduction of rhabdoviruses, in *Comprehensive Virology*, Vol. 4, Fraenkel-Conrat, H. and Wagner, R. R., Eds., Plenum Press, New York, 1975.
21. **Matsumoto, S.**, Rabies virus, *Adv. Virus Res.*, 16, 257, 1970.
22. **Schneider, L. G. and Diringer, H.**, Structure and molecular biology of rabies virus, *Curr. Top. Microbiol. Immunol.*, 75, 153, 1976.
23. **Murphy, F. A.**, Rabies pathogenesis, brief review, *Arch. Virol.*, 54, 279, 1977.
24. **Bradish, C. J. and Kirkham, J. B.**, The morphology of vesicular stomatitis virus (Indiana C) derived from chick embryos or cultures of BHK 21/13 cells, *J. Gen. Microbiol.*, 44, 359, 1966.
25. **Simpson R. W. and Hauser, R. E.**, Structural components of vesicular stomatitis virus, *Virology*, 29, 654, 1966.
26. **Nakai, T. and Howatson, A. F.**, The fine structure of vesicular stomatitis virus, *Virology*, 35, 268, 1968.
27. **Cartwright, B., Smale, C. J., Brown, F., and Hull, R.**, Model for vesicular stomatitis virus, *J. Virol.*, 10, 256, 1972.
28. **Vernon, S. K., Neurath, A. R., and Rubin, B. A.**, Electron microscopic studies on the structure of rabies virus, *J. Ultrastruct. Res.*, 41, 29, 1972.
29. **Cartwright, B., Smale, C. J., and Brown, F.**, Surface structure of vesicular stomatitis virus, *J. Gen. Virol.*, 5, 1, 1969.
30. **Neurath, A.R., Vernon, S. K., Wiener, F. P., Hartzell, R. W., and Rubin, B. A.**, The rabies virus glycoprotein: partial purification and some properties, *Microbios*, 7, 7, 1973.
31. **Peters, D. and Schultz, M. G.**, A model for rhabdovirus morphogenesis, *Proc. K. Ned. Akad. Wet.* 78, 172, 1975.
32. **Brown, F., Smale, C. J., and Horzinek, M. C.**, Lipid and protein organization in vesicular stomatitis and Sindbis virus, *J. Gen. Virol.*, 22, 455, 1974.

33. **Zajac, B. A. and Hummeler, K.**, Morphogenesis of the nucleoprotein of vesicular stomatitis virus, *J. Virol.*, 6, 243, 1970.
34. **Sokol, F., Schlumberger, H. D., Wiktor, T. J., and Koprowski, H.**, Biochemical and biophysical studies on the nucleocapsid and on the RNA of rabies virus, *Virology*, 38, 651, 1969.
35. **Murphy, F. A. and Fields, B. N.**, Kern Canyon virus: electron microscopic and immunlogical studies, *Virology*, 33, 625, 1967.
36. **Murphy, F. A., Shope, R. E. Metselaar, D., and Simpson, D. I. H.**, Characterization of Mount Elgon bat virus, a new member of the rhabdovirus group, *Virology*, 40, 288, 1970.
37. **Kemp, G. E., Lee, V. H., Moore, D. L., Shope, R. E., Causey, O. R., and Murphy, F. A.**, Kotonkan, a new rhabdovirus related to Mokola virus of the rabies serogroup , *Am. J. Epidemiol.*, 98, 43, 1973.
38. **Bauer, S. P. and Murphy, F. A.**, Relationship of two arthopod-borne rhabdovirus (Kotonkan and Obodhiang) to the rabies serogroup, *Infect. Immun.*, 12, 1157, 1975.
39. **Lecatsas, G., Theodoridis, A., and Erasmus, B. J.**, Electron microscopic studies on bovine ephemeral fever virus, *Arch. Gesamte Virusforsch.* 28, 390, 1969.
40. **Ito, Y., Tanaka, Y., Inaba, Y., and Omori, T.**, Electron microscopic observations of bovine epizootic fever virus, *Natl. Inst. Anim. Health Q.*, 9, 35, 1969.
41. **Holmes, I. H. and Doherty, R. L.**, Morphology and development of bovine ephemeral fever virus, *J. Virol.*, 5, 91, 1970.
42. **Murphy, F. A., Taylor, W. P., Mims, C. A., and Whitfield, S. G.**, Bovine ephemeral fever virus in cell culture and mice, *Arch. Gesamte Virusforsch.*, 38, 234, 1972.
43. **Hackett, A. J.**, A possible morphologic basis for the autointerference phenomenon in vesicular stomatitis virus, *Virology*, 24, 51, 1964.
44. **Huang, A. S., Greenwalt, J. W., and Wagner, R. R.**, Defective T particles of vesicular stomatitis virus. I. Preparation, morphology and some biologic properties, *Virology*, 30, 161, 1966.
45. **Crick, J., Cartwright, B., and Brown, F.**, Interfering components of vesicular stomatitis virus, *Nature (London)*, 211, 1204, 1966.
46. **Reichmann, M. E., Pringle, C. R., and Follett, E. A. C.**, Defective particles in BHK cells infected with temperature-sensitive mutants of vesicular stomatitis virus, *J. Virol.*, 8, 154, 1971.
47. **Matsumoto, S.**, Electron microscopic studies on street rabies virus infected mouse, *Ann. Rep. Inst. Virus Res. Kyoto Univ., (Kyoto Daigaku Virusu Kenkyusho Nenkan Kiyo)*, 6, 57, 1963.
48. **Matsumoto, S.**, Electron microscope studies of rabies virus in mouse brain, *J. Cell Biol.*, 19, 565, 1963.
49. **Bergold, G. H. and Munz, K.**, Ultrastructure of Cocal, Indiana, and New Jersey serotypes of vesicular stomatitis virus, *J. Ultrastruct. Res.*, 17, 233, 1967.
50. **David-West, T. S. and Labzoffsky, N. A.**, Studies on the site of replication of vesicular stomatitis virus, *Arch. Gesamte Virusforsch.*, 24, 30, 1968.
51. **David-West, T. S. and Labzoffsky, N. A.**, Electron microscopic studies on the development of vesicular stomatitis virus, *Arch. Gesamte Virusforsch.*, 23, 105, 1968.
52. **Dal Canto, M. C., Rabinowitz, S. G., and Johnson, T. C.**, In vivo assembly and maturation of vesicular stomatitis virus, *Lab. Invest.*, 35, 515, 1976.
53. **Liu, I. K. M. and Zee, Y. C.**, The pathogenesis of vesicular stomatitis virus, serotype Indiana, in Aedes aegypti mosquitoes. I. Intrathoracic injection, *Am. J. Trop. Med., Hyg.*, 25, 177, 1976.
54. **Coward, J. E., Harter, D. H., Hsu, K. C., and Morgan, C.**, Electron microscopic study of development of vesicular stomatitis virus using ferritin-labelled antibodies, *J. Gen. Virol.*, 13, 27, 1971.
55. **Heine, J. W. and Schnaitman, C. A.**, Fusion of vesicular stomatitis virus with the cytoplasmic membrane of L cells, *J. Virol.*, 3, 619, 1969.
56. **Simpson, R. W., Hauser, R. E., and Dales, S.**, Viropexis of vesicular stomatitis virus by L cells, *Virology*, 17, 285, 1969.
57. **Iwasaki, Y., Wiktor, T. J., and Koprowski, H.**, Early events of rabies virus replication in tissue cultures. An electron microscopic study, *Lab. Invest.*, 28, 142, 1973.
58. **Hummeler, K., Koprowski, H., and Wiktor, T. J.**, Structure and development of rabies virus in tissue culture, *J. Virol.*, 1, 152, 1967.
59. **Zee, Y. C., Hackett, A. J., and Talens, L.**, Vesicular stomatitis virus maturation sites in six different host cells, *J. Gen. Virol.*, 7, 95, 1970.
60. **Matsumoto, S. and Kawai, A.**, Comparative studies on development of rabies virus in different host cells, *Virology*, 39, 449, 1969.
61. **Iwasaki, Y. and Clark, H F.**, Cell to cell transmission of virus in the central nervous system. II. Experimental rabies in mouse, *Lab. Invest.*, 33, 391, 1975.
62. **Iwasaki, Y., Ohtani, S., and Clark, H F.**, Maturation of rabies virus by budding from neuronal cell membrane in suckling mouse brain, *J. Virol.*, 15, 1020, 1975.
63. **Iwasaki, Y. and Clark, H F.**, Rabies virus infection in mouse neuroblastoma cells, *Lab. Invest.*, 36, 578, 1977.

64. **Matsumoto, S.**, Electron microscopy of central nervous system infection, in *The Natural History of Rabies*, Vol. 1, Baer, G. M., Ed., Academic Press, New York, 1975, chap. 12.
65. **Dierks, R. E., Murphy, F. A., and Harrison, A. K.**, Extraneural rabies virus infection, *Am. J. Pathol.*, 54, 251, 1969.
66. **Matsumoto, S. and Yonezawa, T.**, Replication of rabies virus in organized cultures of mammalian neural tissues, *Infect. Immun.*, 3, 606, 1971.
67. **Matsumoto, S., Schneider, L. G., Kawai, A., and Yonezawa, T.**, Further studies on the replication of rabies and rabies-like viruses in organized cultures of mammalian neural tissues, *J. Virol.*, 14, 981, 1974.
68. **Murphy, F. A., Bauer, S. P., Harrison, A. K., and Winn, W. C.**, Comparative pathogenesis of rabies and rabies-like viruses. Viral infection and transit from inoculation site to the central nervous system, *Lab. Invest.*, 28, 361, 1973.
69. **Murphy, F. A., Harrison, A. K., Winn, W. C., and Bauer, S. P.**, Comparative pathogenesis of rabies and rabies-like viruses. Infection of the central nervous system and centrifugal spread of virus to peripheral tissues, *Lab. Invest.*, 29, 1, 1973.
70. **Jenson, A. B., Rabin, E. R., Bentinck, D. C., and Melnick, J. L.**, Rabies neuronitis, *J. Virol.*, 3, 265, 1969.
71. **Murphy, F. A., Harrison, A. K., and Bauer, S. P.**, Experimental vesicular stomatitis virus infection: ultrastructural pathology, *Exp. Mol. Pathol.*, 23, 426, 1975.
72. **Miyamoto, K. and Matsumoto, S.**, The nature of the negri body, *J. Cell Biol.*, 27, 677, 1965.
73. **Miyamoto, K.**, Electron microscopic studies on the negri body, *Ann. Rep. Inst. Virus Res., Kyoto Univ. (Kyoto Daigaku Virusu Kenkyusho Nenkan Kiyo)*, 8, 10, 1965.
74. **Miyamoto, K. and Matsumoto, S.**, Comparative studies between pathogenesis of street and fixed rabies infection, *J. Exp. Med.*, 125, 447, 1967.
75. **Murphy, F. A. and Bauer, S. P.**, Early street rabies virus infection in striated muscle and later progression to the central nervous system, *Intervirology*, 3, 256, 1974.
76. **Bergold, G. H. and Munz, K.**, Characterization of Piry virus, *Arch. Gesamte Virusforsch.*, 31, 152, 1970.
77. **Murphy, F. A., Harrison, A. K., and Whitfield, S. G.**, The infection of cells by bullet-shaped viruses, *Proc. Electron Micros. Soc. Am.*, 27, 372, 1969.
78. **Federer, K. E., Burrows, R., and Brooksby, J. B.**, Vesicular stomatitis virus — the relationship between some strains of the Indiana serotype, *Res. Vet. Sci.*, 8, 103, 1967.
79. **Shope, R. E., Murphy, F. A., Harrison, A. K., Causey, O. R., Kemp, G. E., Simpson, D. I. H., and Moore, D. L.**, Two African viruses serologically and morphologically related to rabies virus, *J. Virol.*, 6, 690, 1970.
80. **Tignor, G. H., Murphy, F. A., Clark, H F., Shope, R. E., Madore, P., Bauer, S. P., Buckley, S. M., and Meredith, C. D.**, Duvenhage virus: morphologic, biochemical, histopathologic and antigenic relationships to the rabies serogroup, *J. Gen. Virol.*, 37, 595, 1977.
81. **Garcia-Tamayo, J., Avila-Mayor, A., and Anzola-Perez, E.**, Rabies virus neuronitis in humans, *Arch. Pathol.*, 94, 11, 1972.
82. **Sung, J. H., Hayano, M., Mastri, A. R., and Okagaki, T.**, A case of human rabies and ultrastructure of the negri body, *J. Neuropathol. Exp. Neurol.*, 35, 541, 1976.
83. **Vallat, J. M., Vital, C., Amouretti, M., Faucher, P., and Julien, J.**, Etude clinique et ultrastructurale d'un cas de rage humaine, *Rev. Neurol.*, 133, 637, 1977.
84. **Murphy, F. A., Johnson, H. N., Harrison, A. K., and Shope, R. E.**, Ultrastructural characterization of a new member of the rhabdovirus group — Klamath virus, *Arch. Gesamte Virusforsch.*, 37, 323, 1972.
85. **Murphy, F. A., Coleman, P. H., and Whitfield, S. G.**, Electron microscopic observations of Flanders virus, *Virology*, 30, 315, 1966.
86. **Jenson, A. B., Rabin, E. R., Wende, R. D., and Melnick, J. L.**, A comparative light and electron microscopic study of rabies and Hart Part virus encephalitis, *Exp. Mol. Pathol.*, 7, 1, 1967.
87. **Karabatsos, N., Lipman, M. B., Garrison, M. S., and Mongillo, C. A.**, The morphology, morphogenesis, and serological characterization of the rhabdoviruses Navarro, Kwatta, and Mossuril, *J. Gen. Virol.*, 21, 429, 1973.
88. **Ritter, D. G., Calisher, C. H., Muth, D. J., Shope, R. E., Murphy F. A., and Whitfield, S. G.**, New Minto virus: a new rhabdovirus from ticks in Alaska, *Can. J. Microbiol.*, 24, 422, 1978.
89. **Murphy, F. A. and Shope, R. E.**, unpublished.
90. **Murphy, F. A., Brown, F., and Crick, J.**, unpublished observations.
91. **Monath, T. P., Crop, C. B., and Murphy, F. A.**, personal communication.
92. **Murphy, F. A. and Baer, G. M.**, unpublished.

Chapter 5

THE LIPID ENVELOPE AND CHEMICAL COMPOSITION OF RHABDOVIRUSES

James J. McSharry

TABLE OF CONTENTS

I. INTRODUCTION

The rhabdovirus group encompasses those viruses that have a bullet-shaped or bacilliform morphology. Viruses of plants, insects, fish, and mammals are included in this group. Rhabdoviruses contain one piece of single-stranded RNA which is complementary to messenger-RNA and is enclosed in a helical nucleocapsid complex composed of one or more proteins. The nucleocapsid complex is surrounded by a lipoprotein envelope composed of a lipid bilayer and the envelope proteins. These viruses have been reported to contain 0.7 to 5% RNA, 20 to 25% lipid, 60 to 70% protein, and 3 to 13% carbohydrate.[1-6] The biology of the rhabdoviruses has been the subject of a number of recent reviews.[7-13]

Although there are many rhabdoviruses, the chemical composition of only a few has been studied. These include the plant rhabdoviruses sowthistle yellow vein virus (SYVV), potato yellow dwarf virus (PYDV), and lettuce necrotic yellows virus (LNYV); the fish rhabdoviruses infectious hematopoietic necrosis (INH) virus, spring viremia of carp (SVC) virus, viral hemorrhagic septicemia (VHS) virus, pike fry virus, and swim bladder inflammation virus; and the mammalian rhabdoviruses rabies virus, the various serotypes of vesicular stomatitis virus (VSV), Piry virus, and Chandipura virus. The remainder of this chapter will be a detailed review of the more recent data concerning the structure of the viral envelope and the chemical composition of these rhabdoviruses.

II. THE VIRAL ENVELOPE

The viral envelope consists of an external glycoprotein which is associated with the surface projections, a lipid bilayer, and an internal membrane protein.[14-18] The detailed physicochemical analysis of the viral envelope has been performed only on VSV-Indiana. However, the protein composition of a number of rhabdoviruses has been reported.

A. Proteins of the Viral Envelope

Table 1 lists the proteins associated with the rhabdoviruses of interest to this review. Rhabdoviruses contain four to five major structural proteins.[19] On the basis of the polypeptide profile derived from polyacrylamide gel electrophoresis, rhabdoviruses can be divided into two groups: those with a VSV-like pattern and those with a rabies viruslike pattern. Two viral proteins (G and M) are associated with the viral envelope of VSV-like rhabdoviruses, and three viral proteins (G, M_1, and M_2) are associated with the viral envelope of rabies viruslike rhadoviruses. Three viral proteins (L, N, NS) are associated with the nucleocapsid complex of VSV-like rhabdoviruses, whereas only N has been associated with the nucleocapsid in rabies virus-like rhabdoviruses.

1. The G Protein

The rhabdovirus protein which has a molecular weight ranging between 62 and 85 × 10^3 daltons, incorporates radioactively labeled carbohydrates, and stains with the PAS reagent, has been designated G for glycoprotein.[19] The rhabdovirus glycoprotein is the only viral protein which is exposed to the surface of the virus and is associated with the viral surface projections observed in electron micrographs of negatively-stained virions. The location of the G protein has been determined by selectively removing the peplomers with proteases and determining by polyacrylamide gel electrophoresis which viral proteins are removed from the particle.[14-16,28,31,34] Under these conditions, only the viral glycoprotein was degraded. The smooth-surfaced particles retained their char-

TABLE 1

Proteins of Rhabdoviruses

Virus	L	G	N	NS	M	Ref.
	Molecular weight × 10^3 daltons					
VSV-Indiana	190	69	50	40—45	29	19
	160	65	54	42	27	20
VSV-New Jersey	161	64	52	42	24	20
	—	63	47	39	26	21
VSV-Cocal	—	65	45	38	27	21
Chandipura	162	68	53	—	24	20
Piry	160	75	54	—	23	20
Spring viremia of	150	70	40	—	19	22
carp	>150	82	52	44	23	23
Pike fry rhabdovi-	>150	80	55	—	28	23
rus	160	85	45	—	21	24
Swine bladder inflammation virus	150	83	52	—	25	23
Lettuce necrotic yellows virus	160	75	57	—	22	25
Wheat striate mosaic virus	145	92	59	—	25	6
	L	G	N	M_1	M_2	
Rabies virus ERA strain	190	80	62	40	25	26, 27
HEP — flurry	—	65	58	35	32	28
strain	190	62	54	37	24	29
Infectious haematopoietic necrosis virus	157	72	40	25	20	30
Viral hemorrhagic septicemia virus	157	74	41	22	19	30
Sowthistle yellow	150	83	60	40	36	31
vein virus	—	71	60	42	36	32
Potato yellow dwarf virus	—	78	56	33	22	33

acteristic shape, contained an intact lipid bilayer, and had a lower buoyant density than complete virus. The proteins exposed on the surface of rhabdoviruses have been identified also by surface-specific labeling reagents. The original investigations suggested that G was the major surface protein but that M and possibly L and NS were also exposed to the surface.[35,36] Because of the conflicting results of the protease and surface labeling studies, the problem was reinvestigated using different surface labeling techniques.[37] The results of this study showed that the G protein is the only protein exposed to the virion surface. The earlier results were shown to be due to certain chemical and physical treatments which affected the intactness of the viral membranes. Internal proteins, such as M, are readily iodinated under conditions which lead to disruption of the viral envelope, whereas the nucleocapsid protein is not easily labeled under these same conditions. This labeling pattern led earlier investigators to think that some internal proteins, such as M, L, and NS, were exposed to the surface of the virion. Lactoperoxidase-catalyzed iodination of surface proteins under conditions which insured intact virions results in iodination of only the G protein.[38] Thus, the results of protease experiments and surface labeling experiments have both shown that the VSV-glycoprotein is associated with the surface projections and is the only viral protein exposed to the viral surface. Similar experiments with rabies virus and sowthistle yellow vein virus have shown that the G protein is associated with the surface projections of these viruses also.[28,31]

The G protein of the Indiana serotype of VS contains about 10% carbohydrate by weight and includes mannose, galactose, *N*-acetyl glucosamine, *N*-acetyl neuraminic acid and possibly glucose.[2,4,39,40] The carbohydrate composition can be affected by the host cell in which the virus is grown.[4,39-44] The usual modifications in carbohydrate composition which are determined by the host involve changes in the amount of fucose and sialic acid bound to the terminal portion of the oligosaccharide. The VSV glycoprotein contains two oligosaccharide chains approximately 3400 daltons.[2,4,45,46] A proposal for the structure of the oligosaccharide moiety of the G protein isolated from VSV Indiana propagated in HeLa cells has been published.[45,47,48] The structure was determined by the use of enzymes to sequentially degradate the oligosaccharide, and the preliminary sequence was confirmed by the Smith degradation. Comparison of the structure of the oligosaccharide moieties of the G protein isolated from VSV grown in different cells and from different strains of VSV grown in the same cell will clarify the roles of both the cell and the virus in determining the structure of the carbohydrate side chains of the VSV-glycoprotein.

The glycoprotein isolated from rabies virus grown in BHK21 cells contains 2.9% carbohydrate by weight and consists of mannose, galactose, *N*-acetyl glucosamine, neuraminic acid, and small amounts of fucose.[49] Rabies virus glycoprotein has three different oligosaccharides with molecular weights of 3400, 2800, and 2500 daltons; addition of sialic acid adds heterogeneity to the size of these three oligosaccharide side chains.[50] The carbohydrate composition of the G protein of other rhabdoviruses has not been reported.

In addition to the sugar moieties, the glycoprotein of the Indiana serotype of VSV contains sulfate.[51] When virus is grown in the presence of radioactively-labeled sulfate, only the G protein is radioactively labeled. The ability to label the glycoprotein may be dependent on the cell in which the virus is grown. The G protein of VSV Indiana, grown in HKCC or BHK21-F cells in the presence of radioactively labeled sulfate, is well labeled whereas VSV grown under similar conditions in MDBK[51,52] or CHO cells is poorly labeled.[52] In addition to labeling the VSV-glycoprotein, VSV contains ^{35}S sulfate-labeled material which does not enter the polyacrylamide gel. Pinter and Compans[51] have proposed that this high molecular weight sulfate-containing material is glycosaminoglycan because it is sensitive to testicular hyluronidase and labeled with ^{35}S-sulfate. The function of these virus-associated, sulfate-containing materials are of interest because they may be able to alter virus-host interactions. In addition, the considerable amount of sulfate-containing material may represent some of the host antigens associated with virus particles.[53]

2. The M Protein

The other protein associated with the viral envelope is the membrane (M) protein. This is the smallest protein found in rhabdoviruses, having a molecular weight between 20 and 30 × 10^3 daltons. The membrane protein is thought to be located internal to the lipid bilayer because it is not susceptible to proteolytic digestion or surface labeling in intact virions.[15,16,37,38] The evidence that the M protein is associated with the viral envelope stems from studies involving detergents such as deoxycholate which disrupts many rhabdoviruses into a nucleocapsid complex and an envelope fraction containing G and M proteins. Additional evidence associating the M protein with the viral membrane comes from studies on the synthesis of the M protein. During sythesis the M

TABLE 2

Effect of Cell on Carbohydrate Composition of VS Virions

(% of Total Carbohydrate)

Sugar	L Cell grown Virus	CE-Grown virus
Fucose	2.3	2.5
Mannose	2.9	3.4
Galactose	3.6	0.9
Galactosamine	19.0	13.7
Glucosamine	20.6	17.0
Neuraminic acid	34.3	39.9
Glucose	17.4	22.6

protein is made on free polysomes in the cytoplasm and very rapidly associates with cellular membrane, whereas the proteins of the nucleocapsid complex are synthesized and remain free in the cytoplasm of infected cells.[54] The membrane protein does not associate with the nucleocapsid while it is free in the cytoplasm.[54] The only direct evidence showing the location of the membrane protein comes from studies using protein cross-linkers which showed that the membrane protein can be cross-linked to the nucleocapsid and the glycoprotein, suggesting that it is located between the internal nucleocapsid protein and the surface glycoprotein.[18]

B. Carbohydrate Composition

The original studies on the carbohydrate composition of VSV reported a total carbohydrate content of 13% and showed that both the Indiana and New Jersey serotypes of VSV, when grown in either L-cels or CE fibroblasts, contained fucose, mannose, galactose, glucosamine, and neuraminic acid.[2] There were some quantitative differences in the amount of individual sugars present in the Indiana serotype when it was propagated in L cells or CE fibroblasts, as shown in Table 2. These data suggested that the carbohydrate composition of the viruses was determined by the cell in which it was grown. The carbohydrate composition of the purified glycoprotein isolated from VSV Indiana grown in L cells was similar to that of the whole virus except that it lacked fucose and galactosamine. The absence of these two sugars from the glycoproteins and their presence in the whole virus suggested the presence of other carbohydrate-containing materials in the virion. The presence of hematoside (neuraminosyl-galactosyl-glucosylceramide) in VSV Indiana grown in BHK21-F cells has been reported.[55] Other glycolipids were not identified and quanitated in this study. In addition to glycolipids, VSV contains mucopolysaccharides, which contribute to the overall carbohydrate content of the virion.[51] Recent studies, based on the carbohydrate content of the isolated glycoproteins, reported that the maximum amount of viral carbohydrate of VSV-Indiana grown in BHK21 cells was approximately 3%.[4] The discrepancy between these studies and the original studies is due to a failure of the more recent report to include all of the viral-associated glycolipid and mucopolysaccharide in the calculation. The total carbohydrate composition of other rhabdoviruses has not been reported.

C. Lipid Composition

All rhabdoviruses contain a lipoprotein envelope. The plant rhabdoviruses, potato yellow dwarf virus, lettuce necrotic yellows virus, and wheat striate mosaic virus con-

TABLE 3

Lipid Composition of Rhabdoviruses[1,5]

	VSV Indiana	VSV New Jersey	Rabies virus
Total Lipid	20%	—	24%
Neutral lipid	8	14	6.7
Phospholipid	12	6	15.8
Glycolipid	—	—	1.5
Cholesterol/PL molar ratio	0.64	1.23	0.48
Neutral lipids (Mol %)			
Free fatty acids	3.2	7.6	20.0
Monoglycerides	5.3	4.7	2.4
Diglycerides	10.9	14.6	7.8
Triglycerides	8.6	25.7	5.2
Cholesterol	66.8	37.5	63.4
Cholesterol ester	6.0	9.9	1.2
Phospholipids (Mol %)			
Lysophosphotidylcholine	1.5	3.7	1.5
Sphingomyelin	20.9	22.0	31.3
Phosphotidylcholine	16.4	15.9	23.4
Phosphotidylethanolamine	33.1	31.3	35.1
Phosphotidylinositol/serine	17.3	15.4	8.0
Phosphotidic acid	5.9	4.4	—
Unidentified	4.8	7.3	—

tain 20 to 25% lipid, but further characterization has not been reported.[6,25,56] The buoyant density (1.17 g/cc) of fish rhabdoviruses suggest that they contain lipid, but the lipid composition has not been characterized. However, the Indiana and New Jersey serotypes of VSV and HEP-flurry strain of rabies virus have been studied with regard to their lipid composition. The lipid analyses are presented in Table 3. These rhabdoviruses contain 20% and 24% lipid by weight, respectively.[1,5,57] The major neutral lipid is cholesterol, and the major phospholipids are sphingomyelin and phosphotidyl ethanalamine. The major fatty acids associated with the phospholipids are palimitic, stearic, and oleic acids. The three rhabdoviruses studied showed substantial quantitative differences between the lipid composition of the cellular membranes involved in virus maturation and that of the virus grown in these cells. For example, the plasma membranes of L cells, CE cells, and BHK21 cells contain 30 to 50% phosphotidylcholine whereas the virus grown in these cells contains 16 to 23% phosphotidylcholine; cell membranes contain 30 to 40% cholesterol whereas the viruses grown in these cells contain as much as 63% cholesterol; cell membranes contain 10 to 15% sphingomyelin whereas virus grown in these cells contain twice as much sphingomyelin. There are two possible explanations for the quantitative changes in the lipid composition of these enveloped viruses. First, in order to form the rigid viral membrane which is characteristic of enveloped viruses, the viral proteins may require higher concentrations of cholesterol and sphingomyelin. Secondly, viral infection may alter cellular metabolism in such a way that the synthesis of phosphotidylcholine, which is usually high in cellular membranes and low in viral membrane, is depressed. In actuality, both mechanisms are probably involved in producing the lipid composition of viral envelopes. Studies on the lipid metabolism of virus-infected cells are needed to define the effect of virus infection on the available pool of lipids during virus growth.

It has been postulated that viral proteins can influence the lipid composition of enveloped viruses.[58] This question was addressed by growing VSV Indiana and VSV New

Jersey (which differ in both of their envelope proteins) in the same cell and comparing their lipid compositions.[1,20] There was very little difference between the phospholipid composition of VSV-Indiana and VSV New Jersey grown in either L cells or CE cells. However, there were considerable differences between the composition of the neutral lipid of the two virus serotypes grown in the same cell. In L cell grown virus there was more di- and tri-glyceride and less cholesterol in the VSV New Jersey than in VSV Indiana. A similar situation occurred in the viruses grown in chicken embryo cells. The results of these experiments suggest that viral proteins can affect the selection of neutral lipids in the viral envelope, but have little or no effect on the selection of phospholipids for the viral envelope.

D. Structure of the Viral Envelope

A number of physical techniques including electron spin resononce (ESR), ^{13}C and ^{31}P nuclear magnetic resononce (NMR) and fluorescence polarization have been used to determine the structure of the lipids and proteins of the viral membrane. A description of the physical techniques and their application to the study of the structure of viral envelopes has been published recently.[59] The results of these studies concluded that the viral lipids are arranged in a bilayer and they are more rigid than the lipids in the plasma membranes of uninfected cells used to grow the virus. ESR studies using nitroxide derivatives of stearic acid with the spin label on C_5 or C_{16} to probe the polar and nonpolar portions of the membrane showed that there was a fluidity gradient along the fatty-acid side chain, which exhibited rigidity near the glycerol backbone of the phospholipids and fluidity toward the terminous of the fatty-acid side chains. This result suggests that the lipid was organized in a bilayer structure.[17] Using the C_5 probe to determine the fluidity of the viral membrane compared to the plasma membrane from the cells in which the virus obtained its lipids during virus maturation, ESR studies showed that the viral membrane is more rigid than the plasma membrane. Both the existence of the lipid bilayer and the rigidity of the viral envelope compared to the plasma membrane of cells in which the virus was propagated have been confirmed by other investigators using different techniques to probe the structure of the viral envelope.[60-63]

A number of investigators have attempted to determine what makes the lipid bilayer of the viral envelope more rigid than that of cellular plasma membranes. The results of these experiments suggested that both the lipid-lipid and lipid-protein interactions increased the rigidity of the lipid phase of the viral membrane. The original ESR studies suggested that the viral glycoprotein played a role in determining the rigidity of the polar portions of the bilayer. When virions were treated with protease to remove the exposed portions of the glycoprotein, leaving a small portion of the glycoproteins stuck to the membrane,[34,64] the lipid bilayer became more fluid near the polar region of the membrane probed by the spin label on C_5, but the fluidity at the hydrophobic portion of the membrane probed by the spin label on C_{16} was unchanged.[17] This study suggested that the glycoprotein was adding rigidity to the external surface of the lipid bilayer. These studies were extended using the ^{13}C-NMR technique, which used (N^{13}-CH_3)-choline as a probe for the polar head group and (3-^{13}C)- and (11-^{13}C) oleic acid and (16-^{13}C) palmitic acid to probe the hydrophobic region of the bilayer.[60,61] The result of these studies suggested that the mobility of the polar head groups was restricted by ionic interactions with adjacent phospholipids and/or the glycoproteins, whereas the hydrophobic part of the bilayer was restricted by the high content of cholesterol and possibly interactions with the immersed G peptide and/or M peptide.

Fluorescence polarization measurements of microviscosities of viral membranes and lipid vesicles made fom viral membrane lipids showed similar viscosities indicating that the viral lipids alone attibuted to the rigid structure of the viral envelope.[62] Since the ratio of PE/PC, SpH/PC, and PS/PC are four to five times greater in the viral membrane than in the cellular membrane, they concluded that the ratio of particular phospholipids has a large bearing on the structure of the lipid bilayer. On the other hand, the results of studies using ^{31}P NMR showed that when the glycoprotein was removed by trypsinization, the polar head groups became less rigid. These authors concluded that interactions between the phosphate head groups and the intact glycoprotein add rigidity to the lipid bilayer.[63]

The results of all of these studies suggest that both ionic interactions between the phosphate head group of the adjacent phospholipids and the glycoprotein, as well as the high content of cholesterol, seem to stabilize the lipid in the viral envelope. Since the G protein interacts with only 10% of the viral lipid, lipid-protein interactions with the M protein may contribute more to the rigidity of the viral envelope.[65] Studies of the interaction of envelope lipids with the membrane (M) protein should help elucidate its role in stabilizing the lipid bilayer. Future studies on the structure of the viral envelope of other rhabdoviruses, particularly those of the plant viruses which seem to mature at the nuclear envelope and may contain unusual lipids, should help determine the role of lipids and viral envelope proteins in the structure of the lipid bilayer.

III. THE NUCLEOCAPSID COMPLEX

A. Viral RNA

The RNA isolated from rhabdoviruses is single-stranded and noninfectious and has a sedimentation coefficient between 38 to 45S which corresponds to molecular weights of 3.6 to 4.6×10^6 daltons. A virion-associated, RNA-dependent RNA polymerase capable of converting virion RNA into messenger RNA has been found associated with the nucleocapsid.[12,24,29,66] The structure and function of rhabdovirus RNA has been recently reviewed.[9,12,13]

B. The Proteins of the Nucleocapsid Complex

There are several proteins associated with the nucleocapsid complex. In the case of VSV these proteins are designated L, N, and NS. The major structural protein is the N protein which has a molecular weight of 50,000 daltons. This protein along with virion RNA and two minor proteins (1) L, with a molecular weight of 190,000 daltons and (2) NS, with a molecular weight of 40,000 to 45,000 daltons, form the nucleocapsid complex. This complex is infectious and can transcribe virion RNA into messenger RNA in vivo and in vitro.[12,13] One or more proteins in the nucleocapsid complex may be phosphorylated.[30,67,68]

ACKNOWLEDGMENTS

I would like to thank the following investigators who contributed unpublished data for this review: Frank Landsberger, John Lenard, and Harvey Lodish.

REFERENCES

1. **McSharry, J. J. and Wagner, R.,** Lipid composition of purified vesicular stomatitis viruses, *J. Virol.*, 7, 59, 1971.

2. **McSharry, J. J. and Wagner, R. R.**, Carbohydrate composition of vesicular stomatitis virus, *J. Virol.*, 7, 412, 1971.
3. **Bishop, D. H. L. and Roy, P.**, Dissociation of vesicular stomatitis virus and relation of the virion proteins to viral transcriptase, *J. Virol.*, 10, 234, 1972.
4. **Etchison, J. R. and Holland, J. J.**, Carbohydrate composition of the membrane glycoprotein of vesicular stomatitis virus, *Virology*, 60, 217, 1974.
5. **Blough, H. A., Tiffany, J. M., and Aaslestad, H. B.**, Lipids of rabies virus and BHK-21 cell membranes, *J. Virol.*, 21, 950, 1977.
6. **Trefzger-Stevens, J. M. and Lee, P. E.**, The structural proteins of wheat striate mosaic virus: plant rhabdovirus, *Virology*, 78, 144, 1977.
7. **Howatson, A. F.**, Vesicular stomatitis virus and related viruses, in *Advances in Virus Research*, Vol. 16, Smith, K. M., Louffer, M. A., and Bang, F. B., Eds., Academic Press, New York, 1970, 195.
8. **Matsumato, S.**, Rabies virus, in *Advances in Virus Research*, Vol. 16, Smith, K. M., Louffer, M. A., and Bang, F. B., Eds., Academic Press, New York, 1970, 257.
9. **Knudson, D. L.**, Rhabdoviruses, *J. Gen. Virol.*, 20 (Suppl.), 105, 1973.
10. **Franki, R. I. B.**, Plant rhabdoviruses, in *Advances in Virus Research*, Vol. 18, Louffer, M. A., Bang, F. B., Maramorosch, K., and Smith, K., Eds., Academic Press, New York, 1973, 257.
11. **Printz, P.**, Relationship of sigma virus to vesicular stomatitis virus, in *Advances in Virus Research*, Vol. 18, Louffer, M. A., Bang, F. B., Maramorosch, K., and Smith, K., Eds., Academic Press, New York, 1973, 143.
12. **Wagner, R. R.**, Reproduction of rhabdoviruses, in *Comprehensive Virology*, Vol. 4, Frenkel-Conrat, H. and Wagner, R. R., Eds., Plenum Press, New York, 1975, 1.
13. **Emerson, S. U.**, Vesicular stomatitis virus: structure and function of viral components, *Curr. Top. Microbiol. Immunol.*, 73, 1, 1976.
14. **Cartwright, B., Smale, C. J., and Brown, F.**, Surface structure of vesicular stomatitis, *J. Gen. Virol.*, 5, 1, 1969.
15. **Cartwright, B., Talbot, P., and Brown, F.**, The proteins of biologically active subunits of vesicular stomatitis virus, *J. Gen. Virol.*, 7, 267, 1970.
16. **McSharry, J. J., Compans, R. W., and Choppin, P. W.**, Proteins of vesicular stomatitis virus and of phenotypically mixed vesicular stomatitis virus-simian virus 5 virions, *J. Virol.*, 8, 722, 1971.
17. **Landsberger, F. R. and Compans, R. W.**, Effect of membrane protein on lipid bilayer structure: a spin-label electron spin resononce study of vesicular stomatitis virus, *Biochemistry*, 15, 2356, 1976.
18. **Dubovi, E. J. and Wagner, R. R.**, Spatial relationships of the proteins of vesicular stomatitis virus: induction of reversible oligomers by cleavable protein cross-linkers and oxidation, *J. Virol.*, 22, 500, 1977.
19. **Wagner, R. R., Prevec, L., Brown, F., Summers, D. F., Sokol, F., and MacLeod, R.**, Classification of rhabdovirus proteins: a proposal, *J. Virol.* , 10, 1228, 1972.
20. **Obijeski, J. F., Marchenko, A. T., Bishop, D. H. L., Cann, B. W., and Murphy, F. A.**, Comparative electrophoretic analysis of the virus proteins of four rhabdoviruses, *J. Gen. Virol.*, 16, 1, 1972.
21. **Wunner, W. H. and Pringle, C. R.**, Comparison of the structural polypeptides from vesicular stomatitis virus (Indiana and New Jersey serotypes) and cocal viruses, *J. Gen. Virol.*, 16, 1, 1972.
22. **Lenoir, G.**, Structural proteins of spring viremia of carp, *Biochem. Biophys. Res. Commun.*, 51, 895, 1973.
23. **Hill, B. J., Underwood, B. O., Small, C. J., and Brown, F.**, Physico-chemical and serological characterization of five rhabdoviruses infecting fish, *J. Gen. Virol.*, 27, 369, 1975.
24. **Roy, P., Clark, H. F., Madore, H. P., and Bishop, D. H. L.**, RNA polymerase associated with virions of pike fry rhabdovirus, *J. Virol.*, 15, 338, 1975.
25. **Franki, R. I. B. and Randles, J. W.**, Composition of the plant rhabdovirus lettuce necrotic yellow virus in relation to its biological properties, in *Negative Strand Viruses*, Mahy, B. W. J. and Barry, R. D., Eds., Academic Press, New York, 1975, 224.
26. **Sokol, F., Stancek, D., and Kaprowski, H.**, Structural proteins of rabies virus, *J. Virol.*, 7, 241, 1971.
27. **Madore, H. P. and England, J. M.**, Rabies virus protein synthesis in infected BHK-21 cells, *J. Virol.*, 22, 102, 1977.
28. **Neurath, A. R., Vernon, S. K., Dobkin, M. B., and Rubin, B. A.**, Characterization of subviral components resulting from treatment of rabies virus with tri-(n-butyl) phosphate, *J. Gen. Virol.*, 12, 33, 1972.
29. **Kawai, A.**, Transcriptase activity associated with rabies virion, *J. Virol.*, 24, 826, 1977.
30. **McAllister, P. E. and Wagner, R. R.**, Stuctural proteins of two salmonid rhabdoviruses, *J. Virol.*, 15, 733, 1975.
31. **Ziemiecki, A. and Peters, D.**, The proteins of sowthistle yellow vein virus: characterization and location, *J. Gen. Virol.*, 32, 369, 1976.

32. **Schultz, M. G. and Harrop, K. A.**, The structural polypeptides of sowthistle yellow vein virus (SYVV), *Arch. Virol.*, 50, 173, 1976.
33. **Knudson, D. L. and MacLeod, R.**, The proteins of potato yellow dwarf virus, *Virology*, 47, 285, 1972.
34. **Schloemer, R. H. and Wagner, R. R.**, Association of vesicular stomatitis virus glycoprotein with virion membrane: characterization of the lipophilic tail fragment, *J. Virol.*, 16, 237, 1975.
35. **Walters, G. and Mudd, J. A.**, Iodination of vesicular stomatitis virus with lactoperoxidase, *Virology*, 52, 574, 1973.
36. **Moore, N. F., Kelley, J. M., and Wagner, R. R.**, Envelope proteins of vesicular stomatitis virions: accessibility to iodination, *Virology*, 61, 292, 1974.
37. **Eger, R., Compans, R. W., and Rifkin, D. B.**, The organization of proteins of vesicular stomatitis virions; labeling with pyridoxal phosphate, *Virology*, 66, 610, 1975.
38. **McSharry, J. J.**, The effect of chemical and physical treatments on the lactoperoxidase-catalyzed iodination of vesicular stomatitis virus, *Virology*, 83, 482, 1977.
39. **Etchison, J. R. and Holland, J. J.**, Carbohydrate composition of the membrane glycoproteins of vesicular stomatitis virus grown in four mammalian cell lines, *Proc. Natl. Acad. Sci. U.S.A.*, 71, 4011, 1974.
40. **McSharry, J. J., Ledda, C. A., Freiman, H. J., and Choppin, P. W.**, Biological properties of the VSV glycoprotein. II. Effects of the host and carbohydrate composition on hemagglutination, *Virology*, 84, 183, 1978.
41. **Burge, B. W. and Huang, A. S.**, Comparison of membrane protein glycopeptides of Sindbis virus and vesicular stomatitis virus, *J. Virol.*, 6, 176, 1970.
42. **Moyer, S. A. and Summers, D. F.**, Vesicular stomatitis virus envelope glycoprotein alterations induced by host cell transformation, *Cell*, 2, 63, 1974.
43. **Schloemer, R. H. and Wagner, R. R.**, Mosquito cells infected with vesicular stomatitis virus yield unsialyated virions of low infectivity, *J. Virol.*, 15, 1029, 1975.
44. **Schlesinger, S., Gottlieb, C., Fiel, P., Gelb, N., and Kornfield, S.**, Growth of enveloped RNA viruses in a line of chinese hamster ovary cells with deficient *N*-acetylglucosaminyl transferase activity, *J. Virol.*, 15, 882, 1976.
45. **Robertson, J. S., Etchison, J. R., and Summers, D. F.**, Glycosylation sites of vesicular stomatitis virus glycoprotein, *J. Virol.*, 19, 871, 1976.
46. **Rothman, J. E. and Lodish, H. F.**, Synchronized transmembrane insertion and glycosylation of a nascent membrane protein, *Nature, (London)*, 269, 775, 1977.
47. **Moyer, S. A., Tsang, J. M., Atkinson, P. H., and Summers, D. F.**, Oligosaccharide moieties of the glycoprotein of vesicular stomatitis virus, *J. Virol.*, 18, 167, 1976.
48. **Etchison, J. R., Robertson, J. S., and Summers, D. F.**, Partial structural analysis of the oligosaccharide moieties of the vesicular stomatitis virus glycoprotein by sequential chemical and enzymatic degradation, *Virology*, 78, 375, 1977.
49. **Schlumberger, H. D., Schneider, L. G., Kulas, H. P., and Dieringer, H.**, Gross chemical composition of strain flurry HEP rabies virus, *Z. Naturforsch.*, 28C, 103, 1973.
50. **Dietzschold, B.**, Oligosaccharides of the glycoprotein of rabies virus, *J. Virol.*, 23, 286, 1977.
51. **Pinter, A. and Compans, R. W.**, Sulfated components of enveloped viruses, *J. Virol.*, 16, 859, 1975.
52. **McSharry, J. J.**, unpublished observations, 1978.
53. **Lindenmann, J.**, Host antigens in enveloped RNA viruses, in *Virus Infection and the Cell Surface*, Poste, G. and Nicolson, G. L., Eds., Elsevier/North-Holland Biomedical Press, New York, 1977, 291.
54. **Knipe, D. M., Baltimore, D., and Lodish, H. F.**, Maturation of viral proteins in cells infected with temperature-sensitive mutants of vesicular stomatitis virus, *J. Virol.*, 21, 1149, 1977.
55. **Klenk, H. D. and Choppin, P. W.**, Glycolipid content of vesicular stomatitis virus grown in baby hamster kidney cells, *J. Virol.*, 7, 416, 1971.
56. **Ahmed, M. E., Block, L. M., Perkins, E. G., Walher, B. L., and Kummerviw, F. A.**, Lipids in potato yellow dwarf virus, *Biochem. Biophys. Res. Commun.*, 17, 103, 1964.
57. **Schlesinger, H. R., Wells, H. J., and Hummeler, K.**, Comparison of the lipids of intracellular and extracellular rabies viruses, *J. Virol.*, 12, 1028, 1973.
58. **Blough, H. A. and Tiffany, J. M.**, Theoretical aspects of structure and assembly of viral envelopes, *Curr. Top. Microbiol. Immunol.*, 70, 1, 1975.
59. **Lenard, J.**, Virus envelopes and plasma membranes, *Annu. Rev. Biophys. Bioeng.*, 7, 139, 1978.
60. **Stoffel, W. and Bister, K.**, ^{13}C nuclear magnetic resononce studies on the lipid organization in enveloped virions (vesicular stomatitis virus), *Biochemistry*, 13, 2841, 1975.
61. **Stoffel, W., Bister, K., Schreiber, C., and Tunggal, B.**, ^{13}C-NMR studies of the membrane structure in enveloped virions (vesicular stomatitis virus), *Hoppe-Seyler's Z. Physiol. Chem.*, 357, 905, 1976.

62. **Barenholz, Y., Moore, N. F., and Wagner, R. R.,** Enveloped viruses as model membrane systems: microviscosity of vesicular stomatitis virus and host cell membranes, *Biochemistry*, 15, 3563, 1976.
63. **Moore, N. F., Patzer, E. J., Wagner, R. R., Yeagle, P. L., Hutton, W. C., and Martin, R. B.,** The structure of vesicular stomatitis virus membrane: a phosphorus nuclear magnetic resonance approach, *Biochim. Biophys. Acta*, 464, 234, 1977.
64. **Mudd, H. A.,** Glycoprotein fragment associated with vesicular stomatitis virus after proteolytic digestion, *Virology*, 62, 573, 1974.
65. **Landsberger, F.,** personal communication, 1978.
66. **McAllister, P. E. and Wagner, R. R.,** Virion RNA polzmerases of two salmonid rhabdoviruses, *J. Virol.*, 22, 838, 1977.
67. **Sokol, F. and Clark, H. F.,** Phosphoproteins: structural components of rhabdoviruses, *Virology*, 52, 246, 1973.
68. **Sokol, F., Clark, H. F., Wiktor, T. J., McFalls, M. L., Bishop, D. H. L., and Obijeski, J. F.,** Structural phosphoproteins associated with ten rhabdoviruses, *J. Gen. Virol.*, 24, 433, 1974.

Chapter 6

RHABDOVIRAL RNA STRUCTURE

J. P. Clewley and D. H. L. Bishop

TABLE OF CONTENTS

I. ESTABLISHMENT OF RNA AS THE GENETIC MATERIAL OF VESICULAR STOMATITIS VIRUS

The first studies concerning whether the rhabdovirus vesicular stomatitis virus, (VSV), contains RNA or DNA as its genetic material were those of Cooper[1,2] and Turco.[3] They followed the metabolism of inorganic phosphate (as the radiolabel ^{32}P) in infected cells and showed that small amounts of particulate RNA were liberated from the cells at the same time as infectious progeny virus. In addition, they found that this material sedimented in the ultracentrifuge, and adsorbed to infusioral earth, in a fashion similar to the virus infectivity.[1,2] Further analyses by Turco[3] showed that there was an increase in RNA synthesis at the time of exponential virus release. These results led Cooper to retrospectively suggest that VSV contains RNA as its genetic material.[4]

Direct evidence that VSV contains RNA was obtained by Prevec and Whitmore[5] who purified VSV by differential and equilibrium gradient centrifugation and demonstrated that its nucleic acid component was sensitive to alkali and ribonuclease digestion. Indirect evidence for the nature of the genetic material was also obtained by showing that thymidine analogues, 5-iododeoxyuridine, 5-fluorodeoxyuridine, and 5-bromodeoxyuridine (which are inhibitors of DNA but not RNA synthesis), had little or no effect on plaque formation by VSV when they did reduce the growth of the DNA-containing viruses, vaccinia and adenovirus type 5.[5,6]

Further direct evidence that the virus contains RNA was the demonstration by Huang, Greenawalt, and Wagner[7] that it incorporated ^{3}H-uridine as judged by the coincidence of peaks of radioactivity and infectivity in virus banded on sucrose gradients.

II. PROPERTIES OF THE VIRAL RNA

A. Molecular Weight

A variety of procedures have been used for determining the molecular weight of rhabdoviral RNA. These include sedimentation analyses,[8] length measurements of the viral nucleocapsid,[9] electrophoresis[10] and nuclease digestion.[11] Despite the fact that each of these methods are complicated by technical difficulties which give a degree of inaccuracy to the determined values, there is general agreement that the molecular weights of most, if not all, rhabdoviral RNAs are around 4×10^6. For example, sedimentation measures the average *S* value of the population of molecules so that if the preparation contains a significant proportion of nicked strands, then the true *S* value is underestimated. Polyacrylamide gel electrophoresis is a valuable tool for estimating the apparent molecular weight of an RNA molecule, and a decision on the molecular weight of an unbroken molecule is not influenced by the presence of nicked molecules in the populationn provided that not all are degraded.[12,13] However, it has been shown that certain RNA molecules, when compared to other species, do not exhibit the expected electrophoretic mobility (for example, ribosomal RNA), presumably due to differences in secondary structure and charge effects related to their overall base composition.[12,13] Secondary structure also influences sedimentation constants, as do monovalent and divalent cation concentrations. Thus both sedimentation and electrophoretic mobility determinations can be compromised. Furthermore, both methods rely on the availability of markers of known, and validly determined, molecular weight.

Deduction of the size of the RNA component of the viral nucleocapsid by measurement in the electron microscope[9,14] is an indirect method that is dependent on a number of assumptions; however, it can nevertheless give accurate results.[9]

One procedure developed and used by Repik and Bishop[11] for determining the molecular weight of VSV Indiana RNA involved digesting ^3H nucleoside radiolabeled RNA to completion with base-specific ribonucleases (RNase A and T_1 which give mono- and oligo-adenylate nucleotide sequences terminating in uridine, cytidine, or guanosine). After resolution by DEAE chromatography in 7M urea and computing the recovery of the largest oligonucleotides, the molecular weight of the original RNA was calculated on the basis that after digestion, 1 mol of RNA yielded 1 mol of each particular large oligonucleotide sequence. The technique relies on the availability of sufficient amounts of a pure preparation of RNA (unnicked or partially nicked, provided that all the products are still present), and on nuclease digestion conditions which do not result in significant over- or under-digestion.

1. Sedimentation Coefficient

a. Nondenaturing Conditions

Huang and Wagner[15] showed that ^3H-uridine labeled RNA, released from purified VSV Indiana by detergent lysis, sedimented in a 0 to 15% sucrose gradient with a coefficient of 42 ± 2.4S. From these results the molecular weight of the viral RNA was estimated to be either 4.0×10^6, using the relationship of Spirin:[16]

$$M = 1550\,(S_{20,w})^{2.1} \qquad (1)$$

or 3.1×10^6, using the relationship of Kurland:[17]

$$S = 0.98 \times 10^2\, M^{0.56} \qquad (2)$$

Similar experiments by Brown et al.[18] gave an *S* value of 36 to 40S and a calculated molecular weight of 3×10^6.

The sedimentation coefficient of rabies viral RNA has been reported to be 45S,[19] and its molecular weight, calculated using the relationship of Spirin (see Equation 1), estimated to be 4.6×10^6. Subsequent analyses (see Section II.A.1.) have not confirmed that rabies viral RNA is significantly larger than VSV RNA.[20] Rabies RNA has been shown to have a buoyant density in cesium sulfate of 1.66 g/cm^3.[19]

b. Denaturing Conditions

Mudd and Summers[8] analyzed with known RNA markers the sedimentation coefficient of VSV Indiana viral RNA in both nondenaturing and formaldehyde-denaturing conditions, and found that in nondenaturing conditions the RNA had an *S* value of42.1 ± 1.0S, while in formaldehyde an *S* value of 21.0 ± 0.5S was obtained. The molecular weight of undenatured VSV RNA was estimated by them to be $3.6 \pm 0.1 \times 10^6$ by plotting on a log-log graph the reported *S* values of marker RNAs (Sindbis RNA, polio virus RNA, tobacco mosaic virus RNA, *E. coli* 23S, 16S, and 5S RNA) against their reported molecular weights and correlating the VSV data to that relationship. From the denatured RNA *S* value, the VSV RNA molecular weight was also estimated to be $3.6 \pm 0.2 \times 10^6$ using the relationship of Boedtker:[21]

$$S_{20,w} = 0.05\, M^{0.4} \qquad (3)$$

2. Measurement Studies

Nucleocapsid measurement studies in the electron microscope by Simpson and Hauser[14] were interpreted to show that the viral RNA had a molecular weight of 7×10^6 — a value about twice the true value since they assumed that the ribonucleoprotein

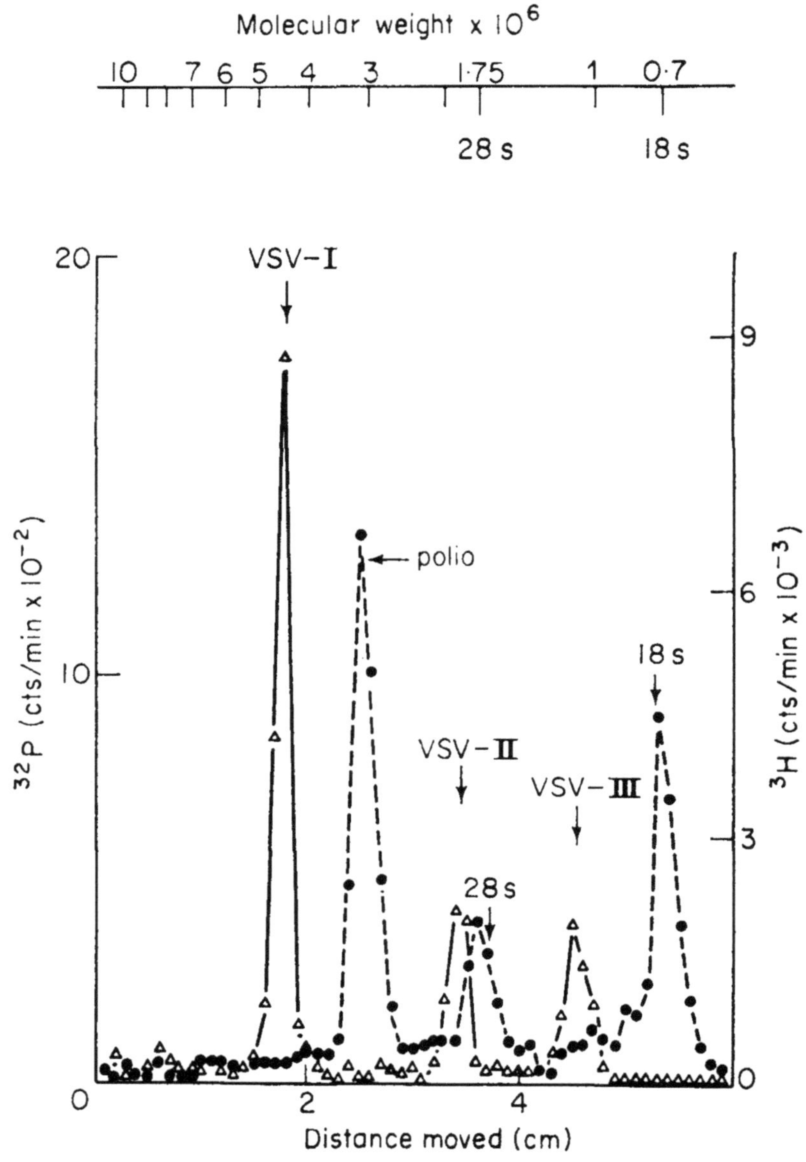

FIGURE 1. Gel electrophoresis of VSV-I, VSV-II, and VSV-III RNA with polio viral, and HeLa ribosomal 28S and 18S RNA, on a 2.2% polyacrylamide gel for 3 hr at 50V and 10mA. (From Bishop, D. H. L. and Roy, P., *J. Mol. Biol.*, 57, 513, 1971. With permission.)

existed in the form of two helices, one inside the other. Nakai and Howatson,[9] on the other hand, estimated from electron micrographs of the viral nucleocapsid that the molecular weight of VSV RNA was 3.4 to 3.8 × 10^6.

3. Electrophoretic Mobility

Analyses of VSV Indiana RNA by polyacrylamide gel electrophoresis were first reported by Bishop and associates[10] using HeLa cell 28S and 18S RNA as markers. They estimated its apparent molecular weight to be 4.4 × 10^6 (see Figure 1). Other estimates have given values for VSV Indiana RNA of 3.9 × 10^6,[22] 3.2 × 10^6,[23] and 4.0 × 10^6,[24]

TABLE 1

Base Ratios of Some Rhabdoviruses

Virus	A	C	G	U	Ref.
Rabies HEP[a]	26.3 ± 0.5	23.8 ± 0.9	20.6 ± 1.0	29.3 ± 0.3	25
Rabies HEP[b]	26.4 ± 0.9	22.5 ± 0.8	21.6 ± 1.5	29.5 ± 0.3	25
Rabies ERA[b]	25.5 ± 0.8	21.7 ± 0.8	24.1 ± 1.2	28.7 ± 0.4	25
Rabies PM[b]	26.7 ± 0.6	22.4 ± 1.5	22.8 ± 0.8	28.1 ± 1.4	25
VSV Indiana[a]	26.0 ± 0.8	23.5 ± 0.6	17.2 ± 1.5	33.3 ± 0.1	25
VSV Indiana[b]	27.4 ± 0.9	21.5 ± 0.6	21.0 ± 0.1	30.3 ± 0.1	25
VSV Indiana[b]	25.2 ± 0.5	18.0 ± 0.5	22.4 ± 0.5	34.4 ± 0.5	11
VSV Indiana[c]	27	19	22	31	11
VSV Indiana[b]	29.3 ± 0.3	21.1± 0.3	20.9 ± 0.2	28.7 ± 0.2	18
VSV Indiana[b]	27.0 ± 0.4	21.4 ± 0.3	20.7 ± 0.3	31.0 ± 0.5	26
Pike Fry	24.9	22.5	20.2	32.4	27

[a] Method for nucleotide separation was Dowex-1-formate chromatography.
[b] Method for nucleotide separation was paper electrophoresis.
[c] Calculated from RNA labeled with ^{3}H-cytidine, uridine, and guanosine.

while values of 4.0 × 10^6 have been reported for VSV New Jersey strain 3366 and Cocal viruses,[24] and 4.5 × 10^1 for VSV Indiana strain Panama BT-78 and VSV New Jersey, Ogden strain.[24] In contrast to these reports of different molecular weights for various rhabdovirus RNAs, Bishop and associates[20] found no significant differences in the eletrophoretic mobilities of the RNAs of VSV Indiana, VSV New Jersey, Kern Canyon, Cocal, Chandipura, Mount Elgon bat, Egtved, Mokola, and the rabies strains HEP, ERA, and CVS.

4. *Nuclease Digestion*

Nuclease digestion analyses of ^{3}H nucleoside-labeled VSV Indiana RNA by Repik and Bishop,[11] involving base ratio determinations and computation of the moles of oligonucleotides obtained from 1 mol of RNA (see discussion above), have given a value of 3.82 ± 0.14 × 10^6 for the viral RNA, corresponding to some 11,278 ± 382 nucleotide bases.

B. Base Ratios

The base ratios of rabies, pike fry rhabdovirus, and VSV Indiana obtained by various investigators are given in Table 1 and, as can be seen, all contain a large amount of uridylic acid.[11,18,25-27] The apparent complementarity of cytosine and guanosine is fortuitous. The base ratios obtained from alkali digestion of ^{32}P-labeled RNA might not necessarily reflect the exact nucleotide composition of the RNA, since it is assumed that the intracellular nucleotide pools of each triphosphate possess the same specific activity.[11] Even though alkali hydrolysis will mitigate the effect of such differences, nonrandom nucleotide neighbor frequencies could compromise the results. However, the computed recoveries of the individual bases in the oligonucleotides obtained from ^{3}H nucleoside-labeled virus have confirmed the overall base composition obtained for VSV Indiana.[11]

C. Amount of RNA Present in the Virion

For VSV Indiana the content of RNA per virus particle was estimated to represent some 2.3% of the virus protein by Cartwright et al.[28] Other estimates[29] have given a slightly lower value (1%) equivalent to some 0.7% of the total virus dry weight mass (RNA, protein, carbohydrate, lipid — see Chapter 5).

D. Organization of the RNA

1. Lack of Infectivity

When extracted and freed from other viral components, the viral RNA of VSV Indiana and rabies is not infectious for tissue culture cells or, in the case of rabies, for mice.[15,19]

2. Existence as a Single, Linear, Covalently-Linked, Single-Stranded Polynucleotide Chain

Since in an early study the viral RNA of VSV was found to sediment as a heterogeneous peak in sucrose gradients (25 to 50S), and a heterogeneous profile of ribonuclease-resistant, viral-specific RNA was obtained in infected cells (7to 11S double-stranded RNA species equivalent to duplexes of 11 to 18S single-stranded RNA), it was suggested that the basic unit of the viral RNA was a complex of four 18S molecules or eight 11S molecules.[30] Other later studies, showing, for instance, a single size class of RNA for native, denatured, or self-annealed VSV Indiana RNA[8,20,23] indicated that the viral RNA exists as a single continuous, covalently linked RNA chain, and not as separate or hydrogen-bonded segments. This conclusion has been confirmed by demonstrating the sensitivity of the RNA to single-strand specific ribonucleases,[11] and the fact that the RNA has a unique 5′-triphosphate terminus.[20,31]

3. Complementarity to Messenger RNA

Messenger RNA isolated from infected cell polysomes has been shown to be complementary to the viral genome of VSV.[8,26,32,33] This, together with the evidence that the viral RNA is not infectious,[15] gave rise to the concept[34] that VSV RNA (negative strand) is opposite in sense to its messenger RNA (plus strand), and led Baltimore et al.[35] to search for, and successfully find, a virion RNA polymerase responsible for transcribing mRNA from the viral genome.

4. Absence of Homopolymeric Tracts of Nucleotides

Under conditions in which polyadenylate sequences could be detected in VSV messenger RNA, no such sequences were found for VSV virion RNA,[36,37] nor were any polyuridylic tracts detected.[38] Oligonucleotide fingerprint analyses of VSV strains also show no evidence for any long homopolymeric tracts of adenosine, uridine, or cytidine.[39]

5. 5′ and 3′ Terminal Nucleotide Sequences

The 5′ nucleotide sequence of VSV Indiana has been shown to be pppApCpGp. . . [31,40] This was deduced from analyses involving (1) alkali digestion of ^{32}P-labeled RNA and chromatography of the products with appropriate standards to identify the 5′ terminal nucleotide as pppAp . . . ; (2) digestion of ^{3}H-adenosine labeled RNA with ribonuclease A and recovery of pppApPyp . . . , followed by identification of the pyrimidine as CMP to give the sequence pppApCp . . . ; and (3) digeston of ^{32}P-labeled RNA with ribonuclease T_1 and recovery of labeled nucleotide tetraphosphate from oligonucleotides expected to contain pppApCpGp . . . .

The 5′ triphosphate terminus of the viral RNA has phosphatase-sensitive phosphates and is therefore uncapped,[31] also there is no evidence that any of the 5′ terminal bases are methylated.[40] However, the 5′ terminal pppApCpGp . . . sequence of the RNA, when it is associated with the nucleocapsid (RNA plus N protein), is protected from ribonuclease A digestion, indicating that the N protein blocks nuclease access to the 5′ terminus of the viral RNA.[31] Likewise, the whole viral RNA sequence in its association with the nucleocapsid is also resistant to nuclease digestion, indicating that N protein protects the phosphodiester backbone of the RNA.[19]

The 3′ terminal sequence was determined by Banerjee and Rhodes[41] to be ... $pPypGpU_{OH}$ by labeling the 3′ end with ^3H-borohydride following oxidation of the RNA. Digestion of such labeled RNA with ribonuclease T_2 or T_1 led to the recovery of label in uridine trialcohol thus identifying the 3′ terminal nucleotide as pU_{OH}, and its neighbor as Gp (i.e., $pGpU_{OH}$); while digestion with ribonuclease A indicated that the next nucleotide was a pyrimidine (Py), giving the deduced sequence of ... $pPypGpU_{OH}$.

It is not known how far the complementarity of the 3′ and 5′ terminal sequences extends, or if it has any circularizing significance as opposed to reflecting a requirement for particlar sequences for RNA replication. Further sequencing will reveal how far the complementarity extends and this will help answer the question of its significance.

6. Coding Capacity

The 11,000 or so nucleotides of the viral RNA[11] could code for around 3700 amino acids (i.e., about 370,000 daltons of protein). The reported molecular weights of the five viral proteins (see review by Bishop and Smith[42]) involve some 330,000 to 350,000 daltons of protein which is equivalent to some 90% of the coding capacity. As the 3′ terminus of the viral RNA is not complementary to an initiator codon, and it is not complementary to the 5′ terminal sequence (m^7GpppmA(m)p(m)A-(m)pCpApGp . . . ...[43-45] of the N messenger RNA (the 3′ proximal gene), it is probable that the 3′ end of the viral RNA does not serve as in a coding capacity for polypeptides. Banerjee and associates[46] have suggested that some 70 bases (complementary to the so-called "leader" RNA) are present at the 3′ end of the viral RNA which, although made in vitro, are not mRNA sequences. In addition, Rose has shown that the N mRNA species (the nearest gene to the viral 3′ terminus) lacks a ribosome binding site at its 5′ terminus,[45] suggesting that there is yet another stretch of bases near the 3′ end of the viral RNA that does not code for protein. Whether mRNA transcription begins with leader transcription is currently being debated[42,46] (as opposed to leader synthesis representing replication initiation).

It is not known if there are intragenic sequences in the viral RNA, or if, at the 5′ end, there is a sequence which is not transcribed into mRNA. Thus the current information suggests that the five viral proteins (N, NS, M, G, and L) are the only proteins coded for by the virus, and that they may well be equivalent to only 90% of the overall nucleotide sequence of the viral RNA.

E. Genome Homologies of Rhabdoviruses

1. Hybridization Studies

Any gross sequence homology that may exist between several rhabdoviruses has been analyzed by hybridizing labeled viral RNA of one virus to an excess of unlabeled viral complementary RNA of another virus and determining the ribonuclease resistance of the hybrids after digesting in nonstringent conditions with ribonuclease A, or in stringent conditions with ribonucleases A and T_1.[20,47,48] For controls, labeled viral RNA was self-annealed or annealed to uninfected, or to homologous virus infected cell nucleic acids. Other than SVCV, and to some extent rabies ERA, self-annealing (e.g., annealing labeled viral RNA to itself or to unlabeled cell RNA) gave essentially little or no evidence of complementary sequences in the viral RNA. The SVCV result is unexplained but reproducible,[52] and since only 40S RNA is obtained from SVCV virions, the result may indicate either that complementary sequences exist in the viral RNA or that complementary 40S RNA species are encapsulated into SVCV virions.

The heterologous annealing results indicate that, using stringent conditions of digestion, there is very little homology between the viral RNAs of VSV Indiana, VSV New Jersey, Cocal, Piry, Chandipura, rabies ERA, and SVCV (Table 2). However, some inexact homology was revealed by the nonstringent digestion conditions between VSV Indiana, Cocal, and VSV New Jersey RNAs. Although Mokola and Lagos bat viral RNA species annealed to some extent to rabies infected cell RNA (nonstringent conditions), at least for the latter the homology was not sufficiently strong to resist the stringent digestion conditions (Table 2). A strong sequence homology was detected between the various rabies strains (Table 2). The hybridization data suggest that certain viruses (e.g., the rabies strains) may have had a common ancestor, although for viruses showing only a little homology the results point to the evolutionary divergence that has occurred.[47] Whether certain nucleotide sequences, such as the replicase binding site, and the complementary sequences to the ribosome binding sites have been conserved between the various rhabdoviruses remains to be determined.

Cross-hybridization studies performed with the purified mRNA species and viral RNAs of different subtypes or isolates of VSV New Jersey (Table 3) have revealed that the Concan, Ogden, and Guatemala isolates are more closely related to each other than to the Hazelhurst and Missouri isolates.[39,48] This result suggests that there has been a relatively recent divergence of VSV New Jersey into two subtypes from a common ancestor. When such divergence occurred is not known.

2. Oligonucleotide Fingerprinting

The oligonucleotide fingerprints of several rhabdoviruses have been obtained[39,49-51] by digesting purified ^{32}P-labeled virion RNA with ribonuclease T_1 and resolving the resulting oligonucleotides by two-dimensional polyacrylamide gel electrophoresis. Figure 2 shows the fingerprint of VSV Indiana and a nomenclature system for the large characteristic oligonucleotides. Two reference dye markers were included in the sample and their final positions are indicated (bromophenol blue, top center, and xzlene cyanol FF, lower center, left). The fingerprint is highly reproducible for duplicate analyses of the same RNA preparation or for various RNA preparations of the same virus strain. The 71 oligonucleotides that have been identified (Figure 2B) represent some 10% of the VSV genome.[39] It has not been possible to trace the origin of our strain of VSV Indiana; however, its fingerprint is indistinguishable from the American Type Culture Collection (ATCC) VSV Indiana strain which was derived from W. E. Cotton's original isolate and is the standard, or prototype strain of VSV Indiana. After 36 consecutive clonings of the prototype strain in BHK-21 cell monolayers, we have not detected any changes in the fingerprint of large oligonucleotides of VSV Indiana,[50] which result points to the relative stability of the viral genome during laboratory passage.

The fingerprint of the RNA of the San Juan strain of VSV Indiana (isolated in 1956 from a diseased steer on the San Juan Indian reservation near Espanola, New Mexico) differs from the prototype strain by lacking 11 large oligonucleotides that the prototype strain has (indicated by stars in Figure 3B), and by having 11 additional oligonucleotides (indicated by arrows in Figure 3B). The fingerprint of the RNA of VSV Indiana strain C, Figure 3D, (isolated in 1946 from a diseased Colorado horse) has six new oligonucleotides which are not present in the prototype strain (Figure 3A), and lacks six or seven of the prototype's oligonucleotides. A recent VSV Indiana isolate, made in 1966 from *Aedes* mosquitoes collected at Rancho de Abiquin, New Mexico,

TABLE 2

Genome Homologies of Rhabdoviral RNAs[a] [42, 47]

Infected cell RNA

Viral RNA	Uninfected	Rabies ERA	SVCV	PFR	Chandipura	Cocal	VSV Indiana	VSV New Jersey	Piry
Rabies ERA	14	98(98)	15	19	18	19	19	ND	17
Lagos bat virus	11	27(4)	3(4)	5(3)	ND	5	13(10)	ND	5
Mokola	1	26	0	ND	ND	ND	ND	ND	ND
SVCV	26	29(22)	80	22	22	22	26	21	25
PRF	3(2)	6	5	(100(96)	5	4	3	3	3
Chandipura	7(4)	8	5	5	100(96)	10(4)	15(6)	16(4)	ND
Cocal	11(4)	6	6	2	13(3)	99(97)	36(10)	24(5)	15(7)
VSV Indiana	5(3)	10	3	2	15(6)	22(10)	97(100)	17(5)	13(6)
VSV New Jersey	8(5)	6	12	12	14(4)	14(5)	20(5)	100(100)	16(7)
Piry	5(3)	2	17	3	14(5)	ND	13(5)	10(5)	97(98)

Various Fixed Rabies Strains[a]

Viral RNA species

Cell RNA	ERA	CVS	HEP	LEP	PM
Rabies ERA	98(89)	97(95)	97(100)	100(100)	94(100)

[a] (^{3}H)-Viral RNA was annealed to uninfected or infected cell RNA and the ^{3}H resistance determined by pancreatic ribonuclease (nonstringent conditions). Values in parentheses are the resistancies after ribonuclease T_1 and pancreatic ribonuclease digestion (stringent conditions). ND = not determined.

TABLE 3

Cross-hybridization of Virion RNA and mRNA Among Strains of New Jersey Serotype[a]

	% mRNA Resistant to RNase					
	Ogden		Cocan		Missouri	
	18—13S	30S	12—18S	30S	13—18S	30S
Ogden	99	94	91	80	—	—
Concan	93	92	97	86	33	33
Guatemala	85	87	—	—	29	31
Missouri	25	19	24	16	94	88
Hazelhurst	31	24	—	—	82	77
No RNA	1	5	0	3	1	14
BHK rRNA	2	5	—	—	1	14

[a] See References 26 and 48.

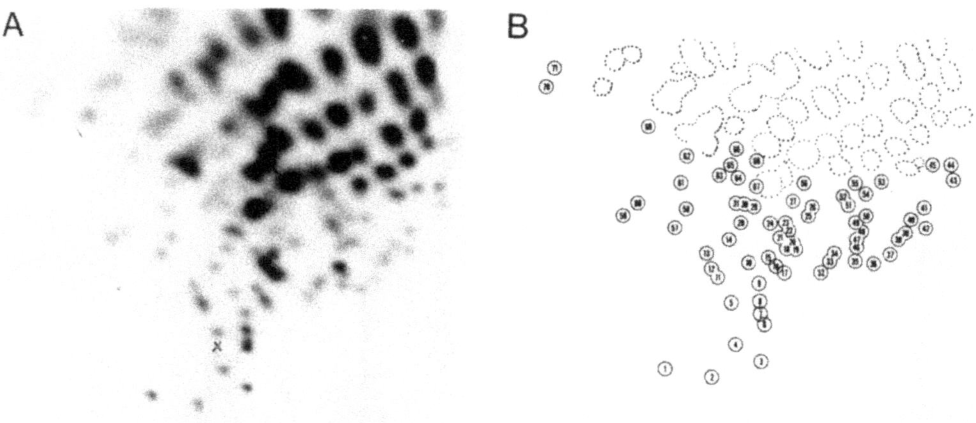

FIGURE 2. The oligonucleotide fingerprint of VSV Indiana B particle RNA (A). A nomenclature system for 71 oligonucleotides resolved in the fingerprint is given in (B). In this and in all subsequent fingerprints, migration in the first dimension gel is from left to right in a 10% polyacrylamide gel, 6 M urea, at pH 3.5. Migration in the second dimension is from bottom to top in a 20% polyacrylamide gel at pH 8.3. The positions of two dye markers are indicated "x": xylene cyanol FF, lower left; bromophenol blue, top right.

(designated VSV Indiana, strain New Mexico) lacks in its fingerprint 11 oligonucleotides by comparison to that of the prototype virus while 8 new oligonucleotides are present (Figure 3C).

The fact that VSV Indiana strains have evolved in different ecological niches can be mimicked in the laboratory by high level mutagenesis.[50] After 23 consecutive passages of VSV Indiana in the presence of 400 μg 5-fluorouracil per mℓ of Eagles' medium, with cloning of the virus in the absence of the mutagen at each passage, 5 of the 71 large oligonucleotides of the prototype virus were sequentially and inheritably lost, while 2 new oligonucleotides appeared (Figure 4). Since the oligonucleotide fingerprint corresponds to some 10% of the nucleotide sequences,[39] these changes probably reflect some 50 nucleotide changes in the genome, representing an overall 0.5% alteration of the genome.

The RNA fingerprints of the five New Jersey strains have confirmed the relatedness

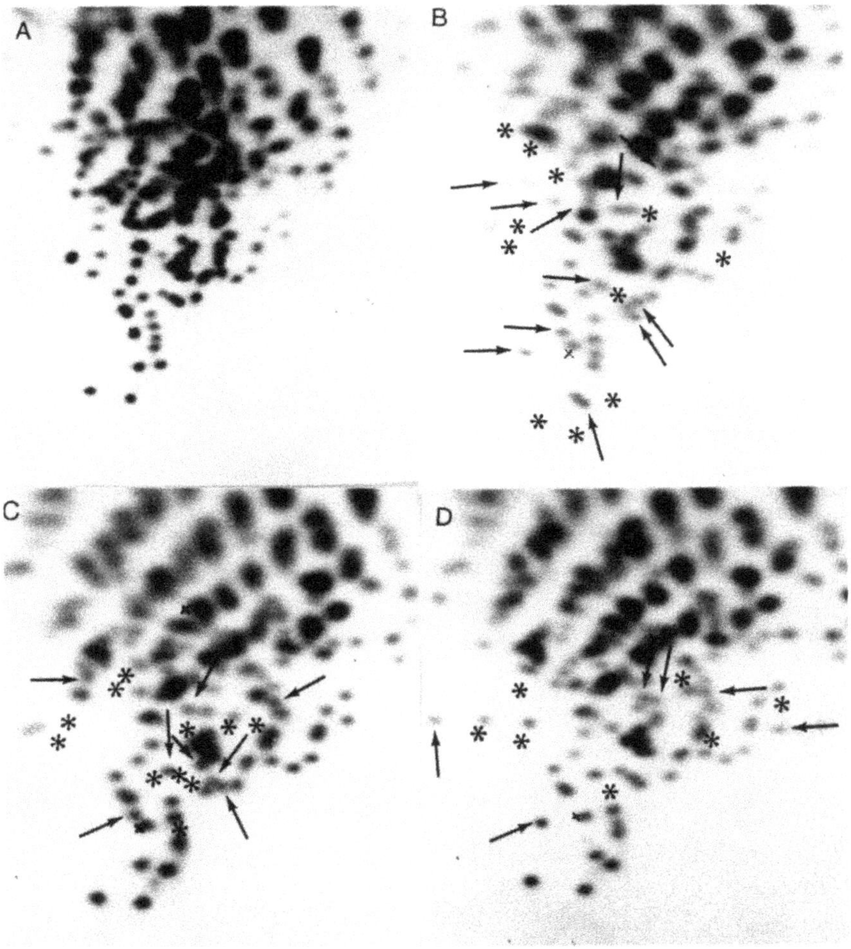

FIGURE 3. The oligonucleotide fingerprints of the B particle RNA of VSV Indiana (Orsay) *ts* O 1-5 (A), VSV Indiana (San Juan) (B), VSV (New Mexico) (C), and VSV Indiana (D) *ts* G I-114. Those oligonucleotides not found in the digest of VSV Indiana (standard) RNA are indicated by arrows, and the oligonucleotides in the VSV Indiana (standard) pattern not found in these patterns are indicated by stars.

of the Concan, Ogden, and Guatemala isolates with each other (Figure 5), and those of the Hazelhurst and Missouri isolates (Figure 6). Also, these analyses have supported the conclusion that there are two VSV New Jersey subtypes.[48]

Although Cocal virus is antigenically, and by RNA homology studies, more closely related to VSV Indiana than to any other virus,[39] its RNA fingerprint is quite distinct from that of the prototype VSV Indiana strain[32] (Figure 7). Likewise, two Chandipura strains, one isolated from a human serum sample obtained in Nagpur, Maharashtra, India, and the other from a hedgehog from Ibadan, Nigeria, have distinct RNA fingerprints (Figure 8). SVCV, as expected from the hybridization data presented previously, shows an RNA fingerprint (Figure 9) that is easily distinguished from those of the other rhabdovirus isolates that have been fingerprinted.[39,49] The RNA fingerprints of rabies strains CVS, ERA, and Kelev are shown in Figure 10.[53] There is apparently more similarity between CVS and Kelev than between CVS and ERA, the significance of which is not known.

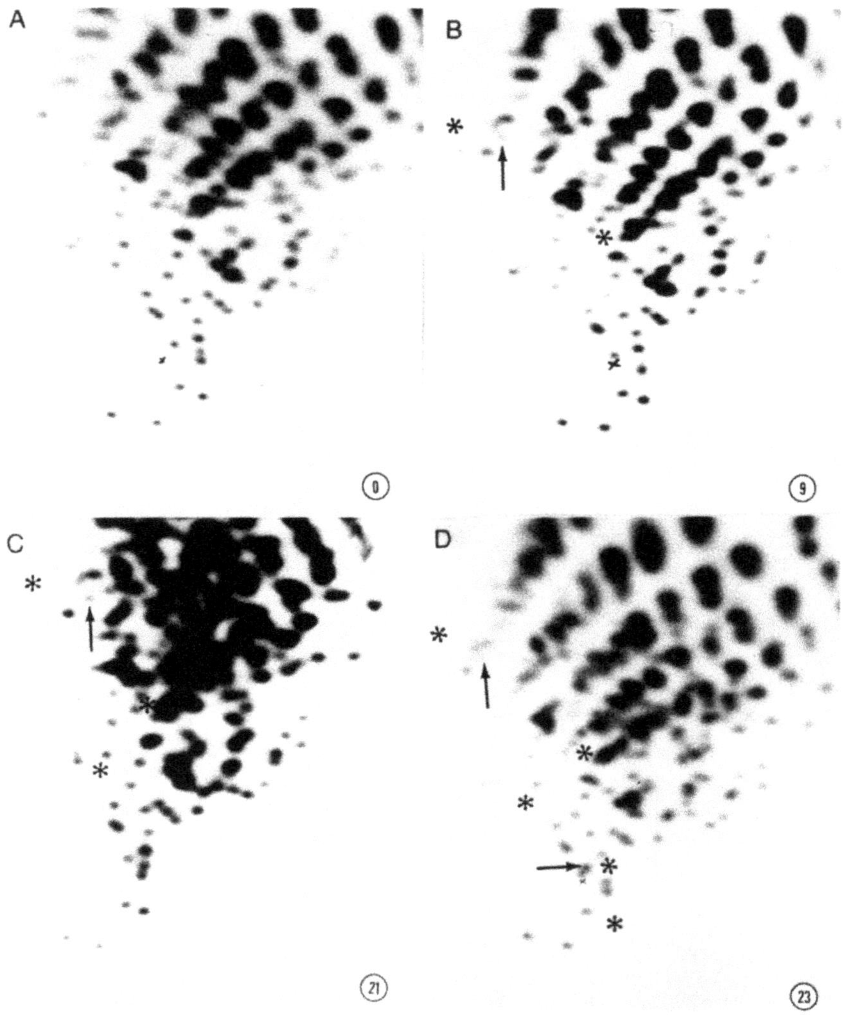

FIGURE 4. The oligonucleotide fingerprints of the RNA of consecutive high level mutagenized VSV Indiana stocks. In (A) the control, passaged virus RNA fingerprint is given. Twenty-three comparative mutageneses were made by growing the virus in the presence of 400 μg 5-fluorouracil per milliliter and cloning the progeny between each mutagenesis. The RNA fingerprints of virus from the 9th (B), 21st (C), and 23rd (D) passage are given.

III. CONCLUSIONS

The RNA of vesicular stomatitis virus represents some 0.7 to 2% of the mass of the virus particle. It is a linear, single-stranded, covalently bonded polynucleotide chain.[8,10,18,23] There is only one copy per virion.[8] Its molecular weight is $3.82 \pm 0.14 \times 10^6$ daltons,[11] equivalent to some 11,000 or so bases,[11] with a base composition of approximately 27% AMP, 21% CMP, 21% GMP, and 31% UMP.[11,18,25-27] The viral RNA sediments with an *S* value of 42S in sucrose gradients.[8,15] Its 5′ sequence is pppApCpGp . . . ,[31] and the terminus is not capped nor are the bases methylated.[40] The 3′-terminus is ...pPypGpU$_{OH}$[41]; it is not polyadenylated.[36,37,39] The viral RNA is not infectious per se[15] and is of a polarity opposite to that of the viral messenger RNA species.[8,26,32,33]

Although the gene order is known (see Chapter 4 Vol. II), how the genome is tran-

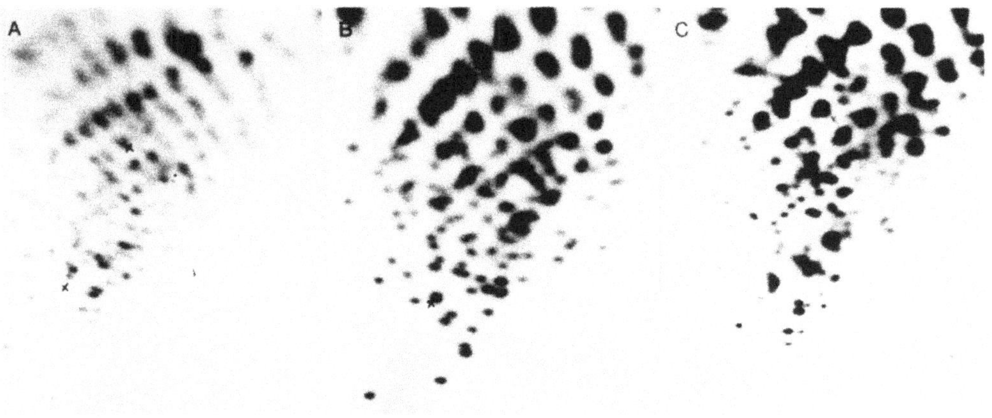

FIGURE 5. The oligonucleotide fingerprints of the viral RNA of three VSV New Jersey strains: (A) Concan, (B) Guatemala, (C) Ogden.

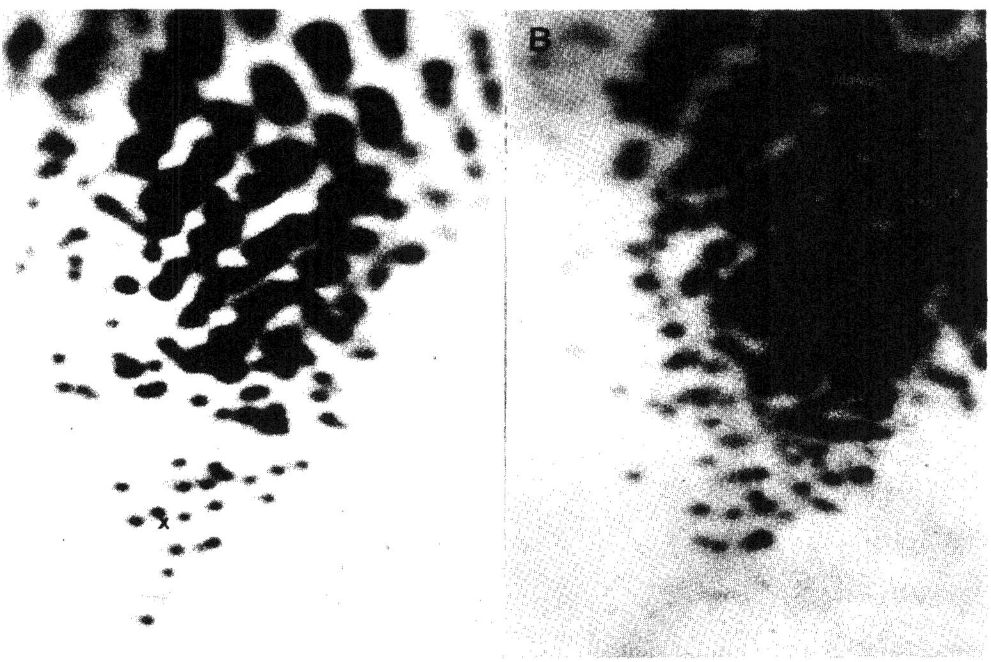

FIGURE 6. The oligonucleotide fingerprints of the viral RNA: (A) VSV New Jersey Hazelhurst, (B) VSV New Jersey Missouri.

scribed is still being debated. The nucleotide sequences proximal to the 5′ and 3′ ends of the genome will undoubtedly be worked out in the near future as, eventually, will be the entire sequence of the RNA, solving many problems pertinent to the organization of the genetic information in the RNA.

Other rhabdoviral RNAs, some of which have been discussed in this chapter and shown to have unique RNA sequences, may well turn out to have properties (e.g., 3′ and 5′ end sequences) which will relate them in specific ways to VSV Indiana even though their gross sequence may be quite different to VSV Indiana (see Chapters 5 to 8 in Vol. III for discussion of other particular rhabdoviruses). Whether under certain

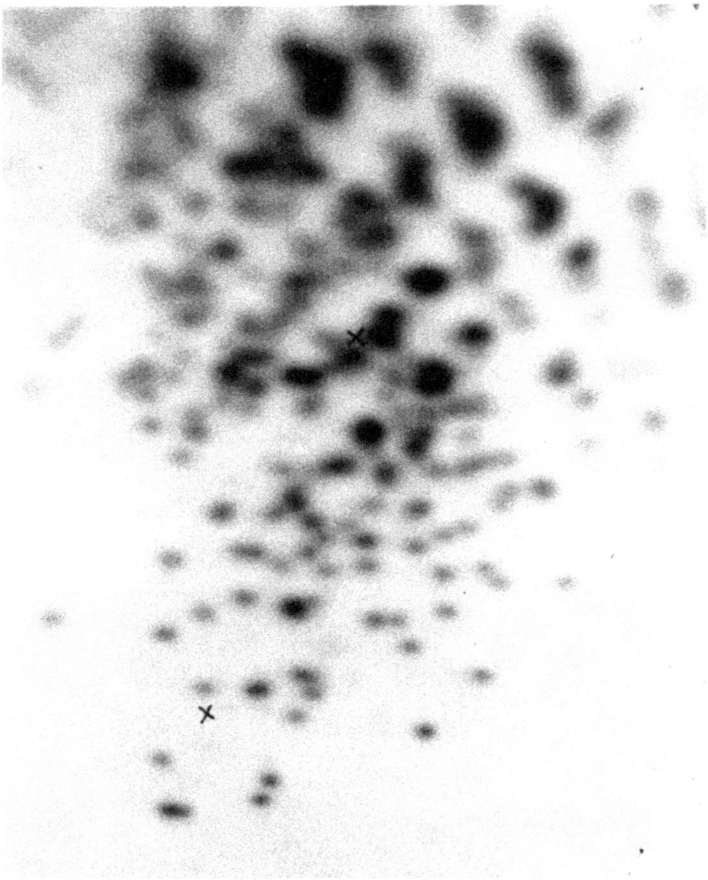

FIGURE 7. The oligonucleotide fingerprint of the particle RNA of Concan virus.

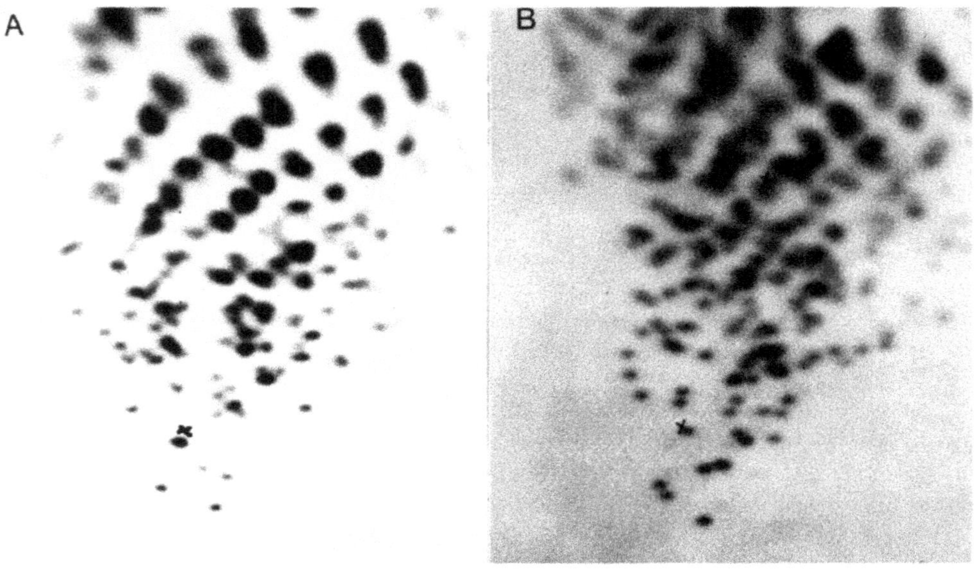

FIGURE 8. The oligonucleotide fingerprint of the particle RNA of two strains of Chandipura virus: (A) fingerprint of Chandipura (Nagpur), (B) fingerprint of Chandipura (Ibadan).

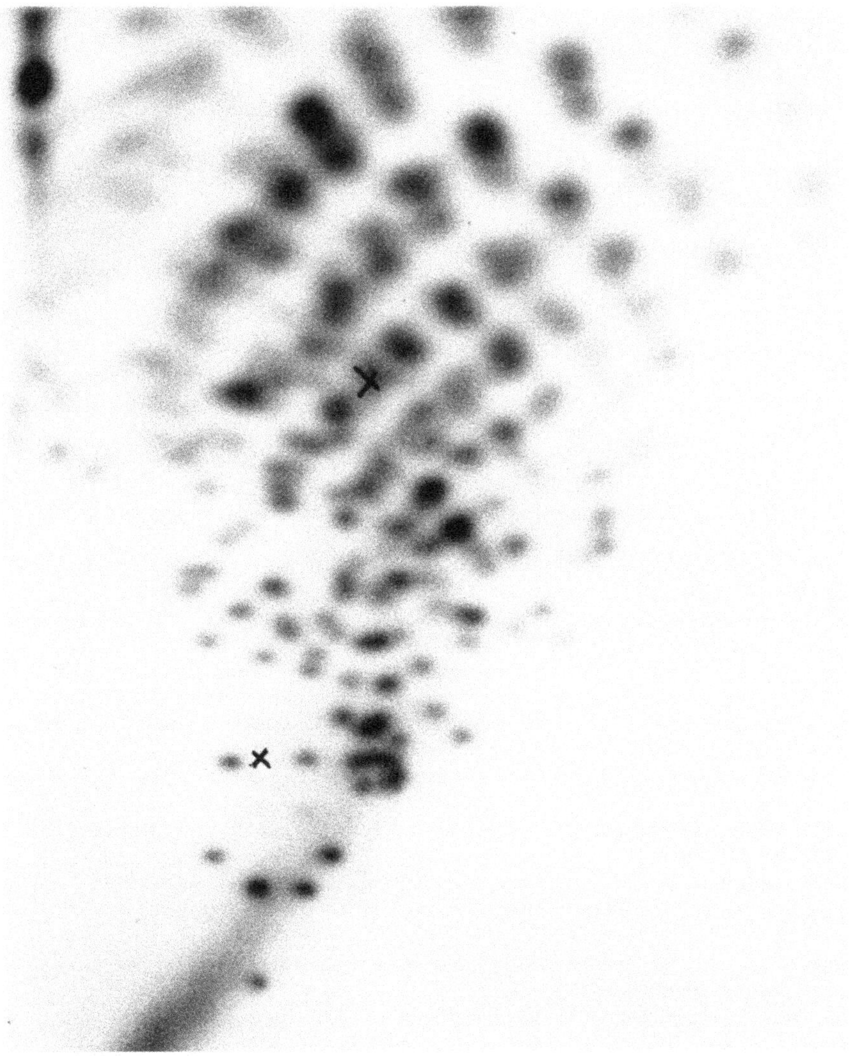

FIGURE 9. The oligonucleotide fingerprint of the particle RNA of spring viremia of carp virus RNA.

FIGURE 10. The oligonucleotide fingerprint of three strains of rabies viral RNA: (A) CVS, (B) Kelev, (C) ERA.

circumstances positive stranded RNA is encapsulated into virions (which the self-annealing data and analyses of certain defective particles suggest — see Chapter 9 Vol. II) needs further study.

REFERENCES

1. **Cooper, P. D.**, Paths of phosphate transfer in normal chick embryo cells and in cells infected with vesicular stomatitis virus, *J. Gen. Microbiol.*, 17, 335, 1957.
2. **Cooper, P. D.**, An osmotic barrier for inorganic phosphate in chick embryo cells and its stability during the latent and release periods of infection by vesicular stomatitis virus, *J. Gen. Microbiol.*, 17, 353, 1957.
3. **Turco, G.**, Changes in cellular ribonucleic acid during growth of vesicular stomatitis virus in chick cell culture, *J. Gen. Microbiol.*, 21, 510, 1959.
4. **Cooper, P. D.**, A chemical basis for the classification of animal viruses, *Nature, (London)*, 190, 302, 1961.
5. **Prevec, L. and Whitmore, G. F.**, Purification of vesicular stomatitis virus and the analysis of P^{32}-labeled viral components, *Virology*, 20, 464, 1963.
6. **Chamsk, H. M. and Cooper, P. D.**, Indirect evidence for the presence of ribonucleic acid in vesicular stomatitis virus, *Virology*, 20, 14, 1963.
7. **Huang, A. S., Greenawalt, J. W., and Wagner, R. R.**, Defective T particles of vesicular stomatitis virus. I. Preparation, morphology and some biologic properties, *Virology*, 30, 161, 1966.
8. **Mudd, J. A. and Summers, D. F.**, Protein synthesis in vesicular stomatitis virus-infected HeLa cells, *Virology*, 42, 328, 1970.
9. **Nakai, T. and Howatson, A. F.**, The fine structure of vesicular stomatitis virus, *Virology*, 35, 268, 1968.
10. **Bishop, D. H. L. and Roy, P.**, Kinetics of RNA synthesis by vesicular stomatitis virus particles *J. Mol. Biol.*, 57, 513, 1971.
11. **Repik, P. and Bishop, D. H. L.**, Determination of the molecular weight of animal RNA viral genomes by nuclease digestions. I. Vesicular stomatitis virus and its defective T particle, *J. Virol.*, 12, 969, 1973.
12. **Loening, U. E.**, The fractionation of high molecular weight ribonucleic acid by polyacrylamide gel electrophoresis, *Biochem. J.*, 102, 251, 1967.
13. **Bishop, D. H. L., Claybrook, J. R., and Spiegelman, S.**, Electrophoretic separation of viral nucleic acids on polyacrylamide gels, *J. Mol. Biol.*, 26, 373, 1967.
14. **Simpson, R. W. and Hauser, R. E.**, Structural components of vesicular stomatitis virus, *Virology*, 29, 654, 1966.
15. **Huang, A. S. and Wagner, R. R.**, Comparative sedimentation coefficients of RNA extracted from plaque-forming and defective particles of vesicular stomatitis virus, *J. Mol. Biol.*, 22, 381, 1966.
16. **Spirin, A. S.**, Some problems concerning the macromolecular structure of ribonucleic acids, in *Progress in Nucleic Acid Research*, Vol. 1, Davidson, J. N. and Cohn, W. E., Eds., Academic Press, New York, 1963, 301.
17. **Kurland, C. G.**, Molecular characterization of ribonucleic acid from *Escherichia coli* ribosomes. I. Isolation and molecular weights, *J. Mol. Biol.*, 2, 83, 1960.
18. **Brown, F., Martin, S. J., Cartwright, B., and Crick, J.**, The ribonucleic acids of the infective and interfering components of vesicular stomatitis virus, *J. Gen. Virol.*, 1, 479, 1967.
19. **Sokol, F., Schlumberger, H. D., Wiktor, T. J., Koprowski, M., and Hummeler, K.**, Biochemical and biophysical studies on the nucleocapsid and the RNA of rabies virus, *Virology*, 38, 65, 1969.
20. **Bishop, D. H. L., Aaslestad, H. G., Clark, H F., Flamand, A., Obijeski, J. F., Repik, P., and Roy, P.**, Evidence for the sequence homology and genome size of rhabdoviral RNAs, in *Negative Strand Viruses*, Vol. 1, Mahy, B. W. J. and Barry, R. D., Eds., Academic Press, London, 1975, 259.
21. **Boedtker, H.**, Dependence of the sedimentation coefficient on molecular weight of RNA after reaction with formaldehyde, *J. Mol. Biol.*, 35, 61, 1968.
22. **Wild, T. F.**, Replication of vesicular stomatitis virus: characterization of the virus-induced RNA, *J. Gen. Virol.*, 13, 295, 1971.
23. **Kiley, M. P. and Wagner, R. R.**, Ribonucleoc acid species of intracellular nucleocapsid and released virions of vesicular stomatitis virus, *J. Virol.*, 10, 244, 1972.

24. **Schaffer, F. L. and Soergel, M. E.**, Molecular weight estimates of vesicular stomatitis virus ribonucleic acids from virions, defective particles, and infected cells, *Arch. Gesamte Virusforsch*, 39, 203, 1972.
25. **Aaslestad, H. G. and Urbano, C.**, Nucleotide composition of the ribonucleic acid of rabies virus, *J. Virol.*, 8, 922, 1971.
26. **Mudd, J. A. and Summers, D. F.**, Polysomal ribonucleic acid of vesicular stomatitis virus-infected HeLa cells, *Virology*, 42, 958, 1970.
27. **Roy, P., Clark, H F., Madore, H. P., and Bishop, D. H. L.**, RNA polymerase associated with virions of pike fry rhabdovirus, *J. Virol.*, 15, 338, 1975.
28. **Cartwright, B., Smale, C. J., Brown, F., and Hull, R.**, Model for vesicular stomatitis virus, *J. Virol.*, 10, 256, 1972.
29. **Bishop, D. H. L. and Roy, P.**, Dissociation of vesicular stomatitis virus and relation of the virion proteins of the viral transcriptase, *J. Virol.*, 10, 234, 1972.
30. **Newman, J. F. E. and Brown, F.**, Induced ribonucleic acids in cells infected with vesicular stomatitis virus, *J. Gen. Virol.*, 5, 305, 1969.
31. **Hefti, E. and Bishop, D. H. L.**, The 5′ sequence of VSV viral RNA and its in vitro transcription product RNA, *Biochem. Biophys. Res. Commun.*, 66, 785, 1975.
32. **Schaffer, F. L., Hackett, A. J., and Soergel, M. E.**, Vesicular stomatitis virus RNA: complementarity between infected cell RNA and RNA's from infectious and autointerfering viral fractions, *Biochem. Biophys. Res. Commun.*, 31, 685, 1968.
33. **Huang, A. S., Baltimore, D., and Stampfer, M.**, Ribonucleic acid synthesis of vesicular stomatitis virus. III. Multiple complementary messenger RNA molecules, *Virology*, 42, 946, 1970.
34. **Baltimore, D.**, Expression of animal virus genomes, *Bacteriol. Rev.*, 35, 235, 1971.
35. **Baltimore, D., Huang, A. S., and Stampfer, M.**, Ribonucleic acid synthesis of vesicular stomatitis virus. II. An RNA polymerase in the virion, *Proc. Natl. Acad. Sci. U.S.A.*, 66, 572, 1970.
36. **Roy, P., Repik, P., Hefti, E., and Bishop, D. H. L.**, Complementary RNA species isolated from vesicular stomatitis (HR strain) defective virions, *J. Virol.*, 11, 915, 1973.
37. **Ehrenfeld, E. and Summers, D. F.**, Adenylate-rich sequences in vesicular stomatitis virus messenger ribonucleic acid, *J. Virol.*, 10, 683, 1972.
38. **Marshall, S. and Gillespie, D.**, Poly U tracts absent from viral RNA, *Nature New Biol.*, 240:43, 1972.
39. **Clewley, J. P., Bishop, D. H. L., Kang, C. -Y., Coffin, J., Schnitzlein, W. M., Reichmann, M. E., and Shope, R. E.**, Oligonucleotide fingerprints of RNA species obtained from rhabdoviruses belonging to the vesicular stomatitis virus subgroup, *J. Virol.*, 23, 152, 1977.
40. **Hefti, E. and Bishop, D. H. L.**, The 5′ nucleotide sequence of vesicular stomatitis viral RNA, *J. Virol.*, 15, 90, 1975.
41. **Banerjee, A. K. and Rhodes, D. P.**, 3′-terminal sequence of vesicular stomatitis virus genome RNA, *Biochem. Biophys. Res. Commun.*, 68, 1387, 1976.
42. **Bishop, D. H. L. and Smith, M. S.**, Rhabdoviruses, in *The Molecular Biology of Animal Viruses*, Nayak, D. P., Ed., Marcel Dekker, New York, 1977, 167.
43. **Rhodes, D. P. and Banerjee, A. K.**, 5′-terminal sequence of vesicular stomatitis virus mRNAs synthesized in vitro, *J. Virol.*, 17, 33 , 1976.
44. **Rose, J. K.**, Heterogeneous 5′ Oterminal structures occur on vesicular stomatitis 'irus mRNAs, *J. Biol. Chem.*, 250, 8098, 1975.
45. **Rose, J. K.**, Nucleotide sequences of ribosome recognition sites in messenger RNAs of vesicular stomatitis in virus, *Proc. Natl. Acad. Sci. U.S.A.*, 74, 3672, 1977.
46. **Banerjee, A. K., Abraham, G., and Colonno, R. J.**, Vesicular stomatitis virus: mode of transcription, *J. Gen. Virol.*, 34, 1, 1977.
47. **Repik, P., Flamand, A., Clark, H F., Obijeski, J. F., Roy, P., and Bishop, D. H. L.**, Detection of homologous RNA sequences among six rhabdovirus genomes, *J. Virol.*, 13, 250, 1974.
48. **Reichmann, M. E., Schnitzlein, W. M., Bishop, D. H. L., Lazzarini, R. A., Beatrice, S. T., and Wagner, R. R.**, Classification of the New Jersey serotype of vesicular stomatitis virus into two subtypes, *J. Virol.*, 25, 446, 1978.
49. **Roy, P. and Clewley, J. P.**, Spring viremia of carp virus RNA and virion-associated transcriptase activity, *J. Virol.*, 25, 912, 1978.
50. **Clewley, J. P. and Bishop, D. H. L.**, Evolution of rhabdovirus genomes, in *Negative Strand Viruses and the Host Cell*, Mahy, B. W. J. and Barry, R. D., Eds., Academic Press, London, in press.
51. **Frisby, D.**, Oligonucleotide mapping of non-radioactive virus and messenger RNAs, *Nucleic Acids Res.*, 4, 2975, 1977.
52. **Roy, P.**, personal communication.
53. **Clewley, J. P., Bishop, D. H. L., and Clark, F.**, unpublished data, 1977.

Chapter 7

RHABDOVIRUS NUCLEOCAPSID

John T. Heyward, Brian P. Holloway, Philip Cohen, and John F. Obijeski

TABLE OF CONTENTS

I. INTRODUCTION

This review is limited to the structure and composition of the rhabdovirus nucleocapsid (NC) and to recent peptide and amino acid analyses of the NC proteins of viruses in the vesicular stomatitis virus (VSV) serogroup.

Rhabdoviruses are pervasive infectious agents of plants and animals and are generally transmitted by arthropod vectors.[1,2] These viruses share common morphological and ultrastructural features and appear as rod-shaped particles 60 to 400 nm long by 60 to 85 nm wide which have an internal core nucleocapsid surrounded by an envelope with protruding spikes. Animal rhabdoviruses are bullet-shaped rods that are flat at one end and tapered at the other, whereas plant rhabdoviruses are usually elongated and tapered at both ends.[1,3-5] Rhabdoviruses are easily adapted to growth in various cell culture systems and are notable in that their growth is inhibited by interferon and for the production of defective-interfering DI particles.[3,6-12]

Rhabdoviruses are divided into subgroups primarily on the basis of antigenic characteristics of the viral surface antigen(s).[2,13-18] Recent studies involving serological cross-reactions of the viral internal proteins revealed further cross-relationships among members of the VSV subgroup, i.e., VSV Indiana, VSV Argentina, VSV Cocal, VSV Brazil, VSV New Jersey, Chandipura, and Piry viruses, and the rabies subgroup including rabies, Mokola, Lagos bat, Kotonkan, and Obodhiang viruses. Conversely, the study of nucleotide sequence homology among the genome RNAs of various rhabdoviruses revealed nearly identical nucleotide sequences among the fixed strains of rabies virus, a very low degree of homology among various members of the VSV serotype, and no homology between rabies virus and either spring viremia of carp virus (SVCV) or VSV.[19-21] In view of these newly discovered interrelationships among members of the VSV subgroup, it will be of interest to examine the biochemical complexity of the nucleocapsid proteins in an effort to correlate these known interelationships with those of protein structure.

II. NUCLEOCAPSID STRUCTURE

Vesicular stomatitis virus may be dissociated by various methods to effect the isolation of different viral components. For example, five virus proteins designated L, G, N, NS, and M by Wagner et al.[22,23] can be resolved by polyacrylamide gel electrophoresis from a preparation of purified VSV treated wih the anionic detergent, sodium dodecyl sulfate (SDS), and a reducing agent. In addition, other minor virion proteins are sometimes found in purified VSV and rabies prepartions.[24,25] The function of these minor components is unknown; however, their apparent molecular weights are similar to that of cellular actin, and they may represent host protein(s) packaged during virus morphogenesis. Rate zonal or equilibrium density gradient centrifugation of VSV treated with sodium deoxycholate or a nonionic detergent (Nonidet P-40®, Triton N-101®, Triton X-100®) releases the virus NC consisting of the virus RNA, the nucleoprotein-N, the large protein-L, and the so-called nonstructural phosphorylated-protein NS.[18,24,26-29] Isopycnic centrifugation of detergent-treated virus in highly concentrated salt such as cesium chloride or in gradients containing sodium chloride at concentrations greater than 0.5 *M* often release the L and NS proteins from the intact NC.[30] An NC structure which contains the L, NS, and N proteins can be recovered easily by centrifugation of Nonidet P-40-treated virus in a hydrated medium of low osmolarity such as sodium diatrizoate.[31] The NC of both VSV and rabies virus has a buoyant density in cesium chloride of 1.32 g/mℓ.[22,26] Sedimentation values of 140 S and 200 S for VSV and rabies virus nucleocapsids, respectively, were obtained by centrifugation of detergent-treated virus preparations in sucrose gradients of low ionic salt.[27,32] The S-values of the various smaller defective interfering particle nucleocapsids would, of course, be less than those obtained for their homologous infectious particles. Intact VSV virion NC can be recovered from the cytoplasm of infected cells which contain the L, N, and NS proteins.[32] Huang and coworkers isolated NC structures from infected cell extracts that contain the nucleoprotein-N and a complementary 42S plus-strand RNA.[33] This structure is probably an important intermediate for the replication of progeny negative-strand virion RNA.

The information gathered by several research groups has revealed that the intact nucleocapsid (i.e., virus RNA and L, N, NS proteins), released from VSV either by a detergent-two phase polymer system or by a nonionic detergent low-salt method, is infectious and capable of complete genome transcription.[24,34-36] Removal of the L and NS proteins from the intact NC by high-salt treatment renders the RNA-N protein complex noninfectious and devoid of any transcriptase activity.[30]

Analysis by electron microscopy of detergent-released VSV nucleocapsid revealed a ribbon-like helically coiled structure with evenly spaced repeating units of N protein along a single strand of RNA. Nakai and Howatson reported the length of completely uncoiled VSV nucleocapsids to average 3.5 μm.[37] Based on this finding they suggested that there was only one NC in each virus particle. The NC isolated from rabies virus has essentially the same morphology and dimensions as VSV nucleocapsids. A more thorough account of the morphology of rhabdoviruses and their nucleocapsids appears in Chapter 4.[38]

III. COMPONENTS OF THE NUCLEOCAPSID

A. Virion RNA

The virion RNAs of all rhabdoviruses tested to date are noninfectious.[5,22,26,27] This observation, coupled with the demonstration that the virus particle contains a virion-associated transcriptase and that complementary messenger RNAs can be obtained from cell polysomes, provides unequivocal evidence that rhabdoviruses are negative-strand viruses.[39-41] Early estimates of the molecular weights of the rhabdoviruses genome ranged from 3.1 to 4.0 × 10^6 daltons and were based on a sedimentation value of the extracted RNA of 42 S.[22] Values, subsequently, derived by coelectrophoresis with appropriate RNA standards were estimated to range from 3.2 to as high as 4.4 × 10^6 daltons.[42,43] Perhaps the most reliable estimate of VSV genome size was provided by Bishop and coworkers, who used a technique involving in vivo labeling of the RNA with specific nucleosides, controlled nuclease digestions of the intact RNA, and separation and quantitation of the resulting oligonuleotides. The value they obtained in this manner was 3.8 ± 0.14 × 10^6 daltons.[20] The genome of rabies virus was calculated to be 4.6 × 10^6 daltons based on a sedimentation coefficient of 45S and a nucleocapsid length of 4.2 μm.[17,21] Recent coelectrophoresis studies with RNAs of VSV and rabies virus in denaturing gels indicated a slightly faster migration rate for VSV RNA.[44] The sizes of RNAs extracted from plant rhabdoviruses are similar to those from animals and range from 4 to 4.7 × 10^6 daltons.[45] Surprisingly, all these values agree well with the estimate of 3.4 to 3.8 × 10^6 daltons calculated by Nakai and Howatson from their measurements of the virus nucleocapsids.[37]

B. L Protein

The largest protein, L, associated with rhabdovirus nucleocapsids has been reported to have a molecular weight ranging from 150,000 to greater than 200,000 daltons by SDS-gel electrophoresis.[46-50] This variation probably reflects the gel electrophoresis system used and the sparsity of reliable molecular weight standards in this size range. Alternative estimates for the size of the L protein can be calculated indirectly from the size of the messenger RNA coding for this protein. In denaturing formamide gels the size of the 28S L-messenger RNA is about 1.65×10^6 daltons or about 4600 nucleotides. A similar value was obtained by size analysis of the duplex RNAs formed between virion RNA and messenger RNA species.[40] From these types of analyses the L protein would have an apparent molecular weight of about 160,000 to 170,000 daltons.[51]

In both VSV and rabies virions the L protein represents about 2 to 4% of the total virus protein and is present in about 30 to 50 copies/virion.[24,52] Due to its large size, early reports suggested that the L protein was an aggregate or precursor of the other virus proteins; subsequently its individuality was established by peptide map analysis of its tryptic and cyanogen bromide peptides.[34,46,53] Several groups have firmly established that the L protein functions as an endogenous RNA-dependent RNA polymerase.[30]

C. N Protein

The nucleoprotein N is the major protein of the intact NC and comprises about 30% of the total virus protein. About 2100 and 1800 copies of the N protein are noncovalently bound to the virion RNA of VSV and rabies virus, respectively.[24,27,29,52] The spatial arrangement of the N protein along the virus RNA probably accounts for the ribonuclease resistance of the isolated NC. Essentially all of the N protein synthesized in VSV-infected cells can be recovered in association with intracellular viral RNA.[32] The function of the N protein in the transcription complex is not known; however, it is essential for transcription, since genome RNA is completely inactive as a template when either L and/or NS protein is added.[30] The N protein of rabies virus and several rabies-like viruses as well as the nucleoprotein of some fish rhabdoviruses are phosphorylated.[21]

D. NS Protein

The NS phosphoprotein of VSV is a minor virion component (about 2%), although it is synthesized in large amounts in infected cells.[25] This protein is found not only in association with intracellular nucleocapsids but also is found free in cell extracts.[32] Recent evidence indicates that both the NS and L proteins are required for the restoration of optimum transcriptase activity to isolated RNA-N protein complexes.[35,36]

Molecular weight estimates of the VSV NS protein obtained with SDS gel electrophoresis vary depending on the gel system used. Initial estimates of about 40,000 to 42,000 were obtained with a neutral pH gel system, whereas estimates of 50,000 or greater were obtained in a Tris-buffered alkaline (pH 8.9) gel system.[49,50,53] This anomalous gel migration of the NS protein is observed whether the protein is from purified virus, from infected cell extracts, or from protein synthesized in a cell-free translation system.[53,54] An alternative method for determining the molecular weight of VSV NS protein is agarose gel filtration in guanidine hydrochloride. With this method, the M and NS proteins coeluted from the gel matrix, indicating that they have very similar molecular weights.[55] An indirect evaluation for the size of the NS protein involved the formation of RNA duplexes between virus and messenger RNA species and the separation of the NS duplex by gel electrophoresis. Subsequent chemical length determinations of the isolated NS messenger RNA species (devoid of poly [A] residues) indicated a size of 2.7×10^5 daltons, which would have a maximum coding capacity of about 31,700.[51] To reconcile these discrepancies in the molecular weight of the NS protein, a more detailed study of the migration rates of phosphoproteins in SDS-gel electrophoresis system is necessary.

TABLE 1

Comparison of Rhabdovirus Nucleocapsid Protein Species. Protein mol wt × 10^3 daltons

Virus	N	L	NS	Ref.
VSV-IND	54	160	42	49
	50	190	53	21
	45	190	50	32
	45	NR[a]	38	50
VSV-NJ	52	161	42	49
	47	NR	39	50
VS-Cocal	45	NR	38	50
Chandipura	53	162	42	56
Piry	54	160	40	56
Rabies	62	NR	NR	21
	54	190	NR	57
Kern Canyon virus (KCV)	54	NR	NR	21
Spring viremia of carp virus (SVCV)	52	150	NR	21
Lagos bat virus	61	NR	NR	21
Mokola virus	64	NR	NR	21
Lettuce necrotic yellows virus (LNYV)	55	100	NR	45
Potato yellow dwarf virus (PYDV)	56	NR	45	58
Obodhiang	52	160	NR	2
Kotonkan	52	150	NR	2
Pike fry rhabdovirus	45	160	NR	59
Infectious hematopoietic necrosis virus (IHNV)	40	157	NR	60
Virus of haemorrhagic septicaemia (VHS)	41	157	NR	60

[a] NR — not reported.

The molecular weights reported for the nucleocapsid-associated proteins of several rhabdoviruses are listed in Table 1.

IV. PEPTIDE MAP ANALYSIS OF VSV NUCLEOPROTEIN

Varying degrees of antigenic relatedness are known to exist among the VSV subgroups of rhabdoviruses, although recent serologic, genetic, and molecular studies revealed considerable divergence among the members. Clewley et al. used a two-dimensional oligonucleotide technique in demonstrating differences in the genome RNAs of VSV-Indiana and VSV-New Jersey serotypes, as well as differences in the RNAs of Indiana serogroup variants.[61] It follows that differences found in the genome RNA must ultimately be reflected in one or more of the viral proteins.

Others have shown by serological methods that changes in the surface glycoprotein G control the separation of VSV strains, whereas, the group-specific antigen was associated with the NC, principally, the N protein.[13-15,18] It therefore seems likely, that in the emergence of new biotypes, the virus surface proteins, particularly the G protein, would vary most. Likewise, the internal proteins associated with the virus RNA would vary least. Doel and Brown have examined this hypothesis by studying the two-dimensional tryptic peptide maps of isolated VSV proteins from several VSV serotypes.[62] They found that the peptides generated from the major internal protein N were very similar, whereas those generated from the G protein varied most.[62] In light of this study and the RNA differences observed by Clewley et al.,[61] we compared the nucleoprotein tryptic peptides of closely related VSV-Indiana strains and VSV-New Jersey by ion-exchange column chomatography. For this study the following viruses were cloned and provided by Dr. D. H. L. Bishop:

1. VSV-Indiana first obtained from the American Type Culture Collection — VSV Indiana ATCC (VSV-IND-ATCC)
2. A temperature-sensitive mutant isolated by Dr. C. Pringle — VSV Indiana *ts*GI-114 (VSV IND *ts*GI-114)
3. VSV strain isolated from an *Aedes* species in New Mexico — VSV-Indiana -New Mexico (VSV IND NM)
4. VSV isolate from a sick animal at the San Juan Indian reservation in New Mexico — VSV Indiana -San Juan (VSV IND SJ)
5. A VSV strain originally isolated from an infected cow in Concan, Texas — VSV New Jersey Concan (VSV NJ CON).

A more detailed account of the isolation and passage history for each of these viruses can be found in a report by Clewley et al.[61]

To determine whether ion exchange chromatography would yield sufficient resolution of tryptic peptide fragments, we examined the chromatograms for ^{35}S-methionine-labeled VSV Indiana structural proteins. Except for the leading peak from the column, there were no apparent areas of the fingerprint which showed significant peptide homology (Figure 1). We reached the same conclusion after examining chromatograms of differentially labeled peptide mixtures of ^{35}S-methionine-labeled L protein with either ^{3}H-methionine labeled G, N, NS, or M protein (data not shown). These observations provide additional evidence that VSV proteins are not derived by cleavage of a precursor protein and suggest that independent translation of monocistronic messenger RNAs specifies each of the structural proteins.[53,54]

The number of ^{35}S-methionine peptides observed in the chromatograms (Figure 1) for the three NC-associated proteins L, N, and NS agreed well with the number found by performing amino acid analysis on the isolated proteins (Table 2). About 11, 13, and 7 methionine peptide peaks were resolved by chromatography for the L, N and NS proteins, respectively, whereas 11, 12, and 6 methionine residues were found for these same proteins by direct amino acid analysis. These apparent minor discrepancies no doubt reflect differences in the methods used.

Having established that ion-exchange chromatography of tryptic-peptides provided sufficient resolution for examining differences among proteins, we compared the peptides of differentially labeled preparations of the N proteins from closely related VSV Indiana strains. To do this the N protein from VSV IND ATCC (prototype virus) labeled with either ^{35}S-methionine or ^{14}C-lysine was mixed with the corresponding ^{3}H-methionine or ^{3}H-lysine labeled N protein from either VSV-IND *ts* GI-114 or VSV IND NM or from VSV IND SJ (Figures 2 and 3). These chromatograms indicated that the N protein peptide sequence of these viruses had been strictly conserved. We had anticipated that minor differences among these proteins might be resolved by analyzing the fragments of lysine labeled mixtures (Figure 3). Although about twice the number of peptides were resolved with this labeling system, the chromatograms were very similar.

To determine whether differences could be found between the N protein of VSV Indiana and VSV New Jersey, differentially labeled mixtures of each were digested with trypsin and applied to the column (Figure 4). Obvious and reproducible differences were apparent in these chromatograms, not only in the number of peptides resolved, but also, in their elution position from the column. Doel and Brown found similar differences for the N protein of these same viruses with a two-dimensional mapping procedure.[62]

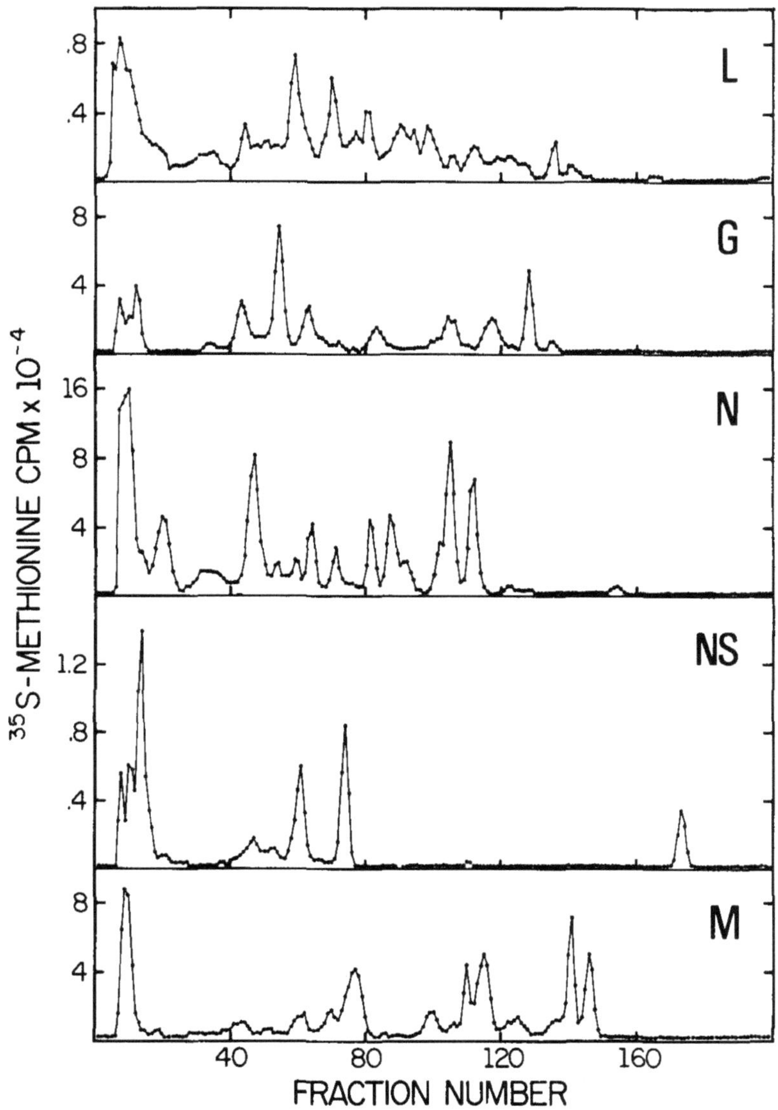

FIGURE 1. Fingerprint of ${}^{35}S$-methionine-labeled VSV-IND-ATCC viral proteins (L, G, N, NS, and M). In this chapter, the peptide fingerprints of VSV proteins were developed as described in the following procedures: (A) Isolation and preparation of the viral proteins; (B) Ion-exchange chromatography of tryptic peptides. (A) Isolation and preparation of VSV viral proteins: Purified virus (1 mg/mℓ in 0.1 *M* sodium phosphate buffer, pH 7.0) was dissociated with a 3% (w/v) mixture of SDS and 2-mercaptoethanol and heated at 100°C for 1 min. Samples containing 3 to 5 × 10^6 cpm of radioactivity (about 40 to 70 μg protein) were applied to cylindrical 8% acrylamide-SDS-phosphate gels, and the proteins were separated by electrophoresis at 3 ma/gel for 16 hr.[49] Protein bands detected in the gel with Coomassie brilliant blue stain were excised and placed in separate vials. To aid the elution of proteins from the acrylamide gel, the bands were first minced with a scalpel and then pulverized to form a slurry by forcing the gel several times through a canula (17G) into a vial containing 1 to 2 mℓ of protein elution buffer (0.02 *M* sodium phosphate, pH 7.0, 0.2% SDS, and 0.5% 2-mercaptoethanol). The slurry was stirred overnight at room temperature, and the acrylamide pulp was separated from the fluid by centrifugation at 11,000 rpm for 20 min and 15°C. The acrylamide pellet was extracted a second time with 2 mℓ of protein elution buffer for 8 hr. The supernates were pooled, and residual acrylamide granules were removed by centrifugation. Each protein was reduced with 250 μg of dithiothreitol for 30' min at 30°C and then carboxymethylated with 500 μg iodoacetamide for 1 hr in the dark. Twenty micrograms of carrier protein (bovine serum albumin) was added to each protein sample, and the proteins were precipitated with three volumes of cold acidic methanol (1% glacial

acetic acid) at −20°C for 16 hr. The proteins were recovered by centrifugation at 25,000 rpm for 30 min at 4°C in a Beckman SW 41 rotor. The supernate was removed, and the pellet was washed twice with methanol and dried. Each protein was dissolved in 1 mℓ of 0.1 *M* ammonium bicarbonate buffer (pH 7.9) and was digested with 10 μg chymotrypsin-free trypsin for 16 hr at 37°C. To insure complete digestion another 10-μg portion of trypsin was added to each protein and digestion was continued an additional 4 hr at 37°C. The digests were lyophilized, dissolved in 1 mℓ of 0.2 *M* pyridine-acetate buffer (pH 3.1) and insoluble material was removed by centriugation. Before chromatography the digest was passed through a 0.22 μ Millex filter. (B) Ion-exchange chromatography of tryptic peptides: An ion exchange column was prepared with cationic Chromobead-P (Technicon).[53] The column was a jacketed glass unit (9 × 240 mm) connected by Teflon® tubing to form a closed chromatographic system including a gradient maker, an adjustable pulseless pump, an automatic 3 mℓ sample injection loop, and a fraction collector that accommodated two hundred 7 mℓ glass scintillation vials. The column was operated at 51°C under a constant pressure of 125 psi with an elution rate of 30 mℓ/hr. The column was developed with a linear gradient of 100 mℓ of leading buffer (0.2 *M* pyridine acetate, pH 3.1) and 100 mℓ of trailing buffer (2 *M* pyridine acetate, pH 5.0). When used, the column was equilibrated in leading buffer, and a 2 mℓ sample was applied via the injection loop. After the first ten 1 mℓ fractions eluted from the column, the concentration-pH gradient was started by allowing an even flow and constant mixing of the trailing buffer (100 mℓ) with the leading buffer (100 mℓ). At the end of a run the fractions were placed in an oven at 70°C and dried. The dried residue was dissolved in 6 mℓ of T-T cocktail (20 parts toluene, 10 parts Triton X-100®, and 3 parts water) and monitored for radioactivity. After each run the column was flushed with 5 vol of trailing buffer followed with 5 vol of the leading buffer.

TABLE 2

Amino Acid Compositions of the L, N, and NS Proteins of VSV-Indiana

	Molar ratios[a]			Residues/mole of protein[b]		
	L	N	NS	L	N	NS
Aspartate	1.004	1.633	1.073	169	63	26
Threonine	0.574	0.878	0.642	97	34	16
Serine	0.991	1.116	1.057	167	43	26
Glutamate	1.029	1.153	1.506	173	45	37
Proline	0.531	0.633	0.497	90	24	12
Glycine	1.143	1.177	1.003	193	46	25
Alanine	0.603	0.973	0.718	102	38	18
Valine	0.531	0.901	0.691	90	35	17
Methionine	0.066	0.307	0.244	11	12	6
Isoleucine	0.614	0.530	0.515	104	21	13
Leucine	1.000	1.000	1.000	169	39	25
Tyrosine	0.302	0.433	0.393	51	17	10
Phenylalanine	0.427	0.405	0.382	72	16	9
Histidine	0.261	0.129	0.222	44	5	5
Lysine	0.564	0.650	0.855	95	25	21
Arginine	0.574	0.487	0.481	96	19	12

[a] VSV L, N, and NS proteins were purified by polyacrylamide gel electrophoresis as described[63] and were hydrolysed for 22 h at 110°C in 6 N — HCl and 0.01M phenol. Amino acids were analysed on a Beckman "Multichrom" two-column system. All values were corrected for a 5% and 10% destruction of threonine and serine, respectively, and by assuming that the release of valine and isoleucine was only 90% complete. Performic acid oxidation of the samples was not done; therefore the amount of cysteine could not be determined.

[b] Assumes molecular weights of the 190,000, 52,500, 31,700 daltons for the L, N, and NS proteins, respectively.

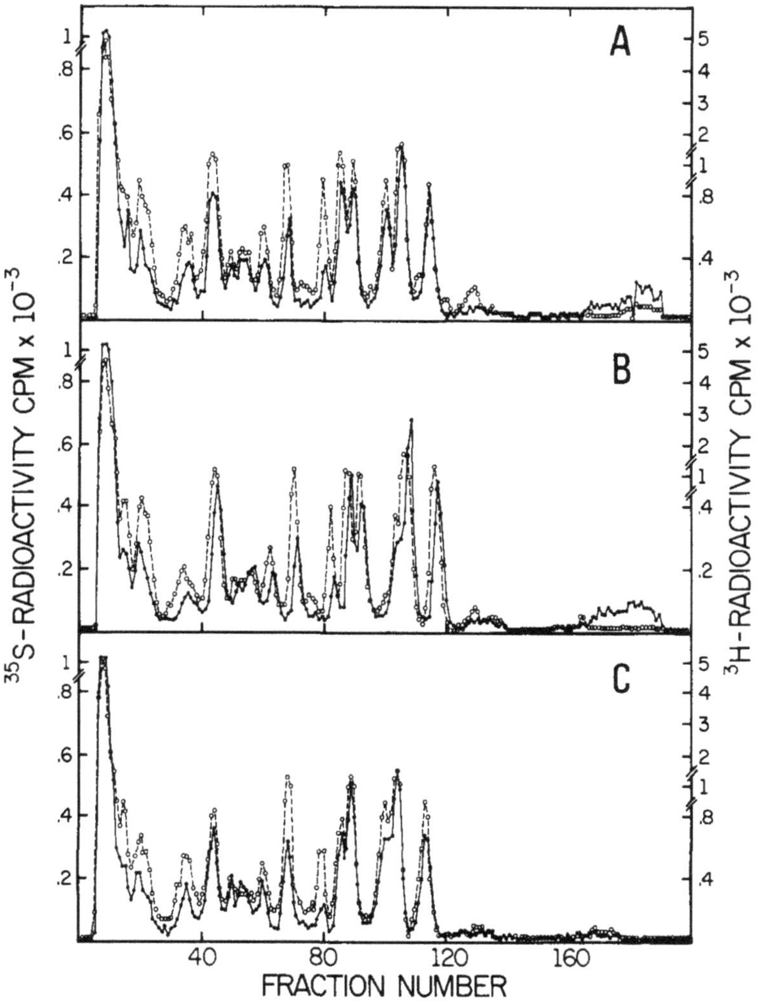

FIGURE 2. Tryptic peptides of ^{35}S-methionine-labeled VSV-IND-ATCC N protein compared to the ^{3}H-methionine N protein peptides of (A) VSV-IND tsGI-114, (B) VSV-IND-NM, and (C) VSV-IND-SJ. Differentially labeled samples of the virus proteins were separated by SDS-gel electrophoresis and the N protein eluted. Mixtures of the VSV-IND-ATCC N protein and the other VSV-IND N protein were prepared for analysis of their tryptic peptides as described. Symbols: •,^{35}S; O,^{3}H.

V. AMINO ACID ANALYSIS OF VSV NUCLEOCAPSID PROTEINS

VSV L, N, and NS proteins were purified by preparative SDS-polyacrylamide gel electrophoresis, and the purity of each protein was monitored by analytical gel electrophoresis. Performic acid oxidation of the samples was not attempted; therefore the amount of cysteine was not determined. The total number of polar amino acids residues (histidine, lysine, arginine, aspartate, threonine, and serine) was about 50% for each protein (Table 2). The total acidic amino acid levels (aspartate and glutamate) were between 2.033 and 2.786; the highest content was found in the N protein. All three proteins contained a relatively small proportion of basic amino acids (histidine + lysine + arginine = 1.266 to 1.558); the lowest level being that of the N protein. The low content of basic amino acids and the high proportion of acidic residues in the major nucleocapsid protein N was not totally unexpected, because others have reported

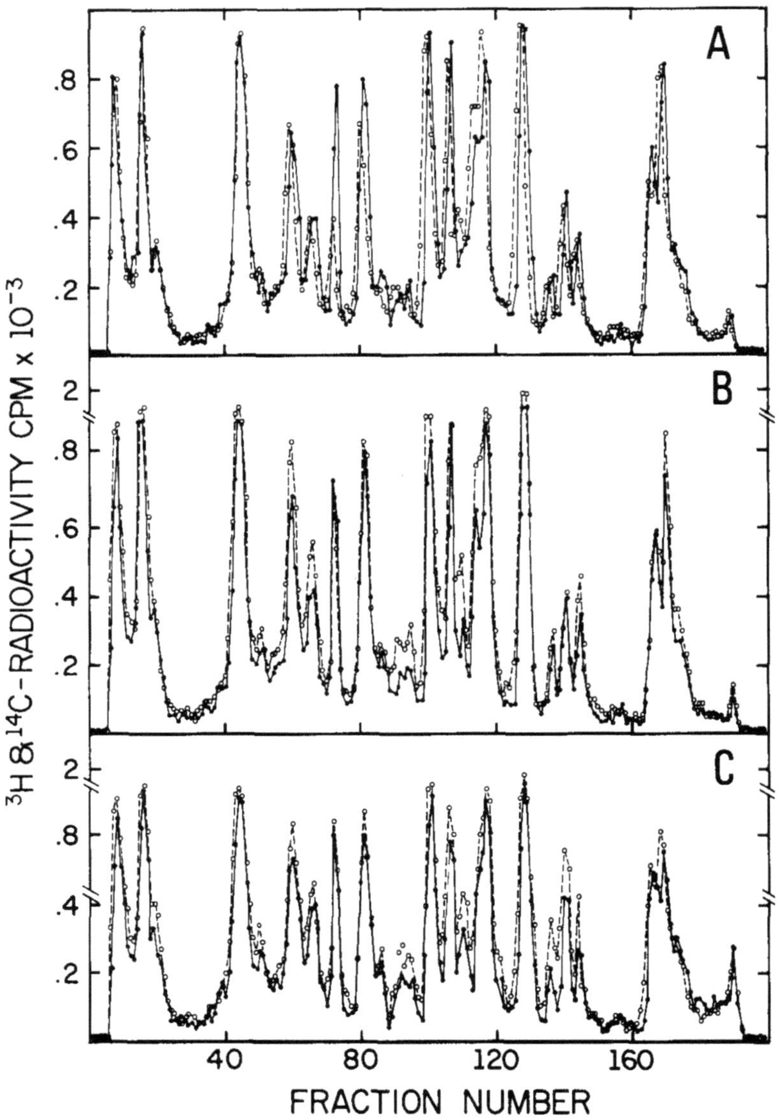

FIGURE 3. Fingerprints of ^{14}C-lysine-labeled VSV-IND-ATCC N protein compared to those of the ^{3}H-lysine-labeled N protein peptides of (A) VSV-INDtsGI-114, (B) VSV-IND-NM, and (C) VSV-IND-SJ. Mixtures of differentially labeled N protein peptides were prepared and separated by chromatography as described. Symbols: •, ^{14}C; O, ^{3}H.

similar results for the nucleocapsid protein of Newcastle disease virus and influenza virus.[63,64] Results of these analyses can be compared with those showing a relatively high basic amino acid content in the nucleocapsid protein of Semliki Forest virus, a member of the Togaviridae.[65] Since the N protein of VSV is tightly bound to the genome RNA, this protein might be expected to contain a high concentration of basic amino acids; however, our results did not support this hypothesis. These findings, therefore, emphasize the need to study the higher-order structure of viral protein-nucleic acid interactions.

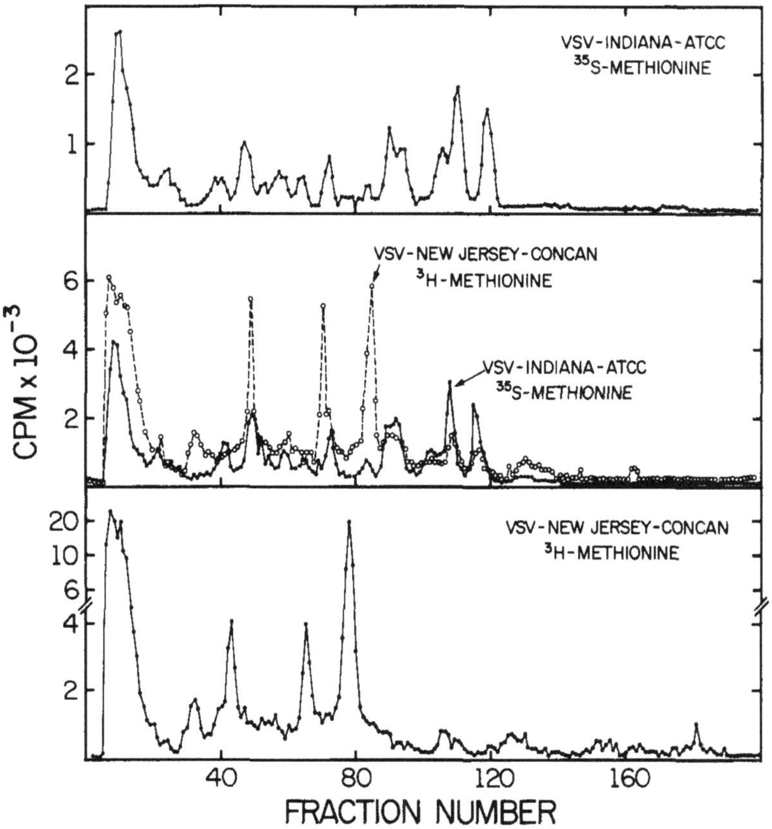

FIGURE 4. Tryptic peptides of differentially labeled N proteins from VSV-IND-ATCC and VSV-NJ-CON. The ^{35}S- and ^{3}H-methionine-labeled N protein of VSV-IND-ATCC and VSV-NJ-CON, respectively, were separated by SDS-gel electrophoresis, eluted from the gel, and prepared for analysis of their tryptic peptides as described. TOP: Tryptic peptides of VSV-IND-ATCC N protein labeled with ^{35}S-methionine. MIDDLE: Mixture of N protein peptides from methionine-labeled VSV-IND-ATCC (•) and ^{3}H-methionine-labeled VSV-NJ-CON (O). BOTTOM: Tryptic peptides of VSV-NJ-CON N protein labeled with ^{3}H-methionine.

REFERENCES

1. **Francki, R. I. B.**, Plant rhabdoviruses, in *Advances in Virus Research*, Vol. 18, Lauffer, M. A., Bang, B. B., Maramorosch, K., and Smith, K., Eds., Academic Press, New York, 1973, 257.
2. **Bishop, D. H. L. and Smith, M. S.**, Rhabdoviruses, in *The Molecular Biology of Animal Viruses*, Vol. 1, Nayak, D. P., Ed., Marcel Dekker, New York, 1977, 167.
3. **Hackett, A. J., Zee, Y. C., Schaffer, F. L., and Talens, L.**, Electron microscopic study of the morphogenesis of vesicular stomatitis virus, *J. Virol.*, 2, 1154, 1968.
4. **Howatson, A. F.**, Vesicular stomatitis and related viruses, *Adv. Virus Res.*, 16, 195, 1970.
5. **Knudson, D. L.**, Rhabdoviruses, *J. Gen. Virol.*, 20, 105, 1973.
6. **McCombs, R. M., Benyesh-Melnick, M., and Brunschwig, J. P.**, Biophysical studies on vesicular stomatitis virus, *J. Bacteriol.*, 91, 803, 1966.
7. **Franklin, R. M.**, Studies on the growth of vesicular stomatitis virus in tissue culture, *Virology*, 5, 408, 1958.

8. **Huang, A. S. and Baltimore, D.**, Defective viral particles and viral disease processes, *Nature (London)*, 226, 325, 1970.
9. **Huang, A. S. and Wagner, R. R.**, Defective T-particles of vesicular stomatitis virus. II. Biological role in homologous interference, *Virology*, 30, 173, 1966.
10. **Schaffer, F. L., Hackett, A. J., and Soergel, M. E.**, Vesicular stomatitis virus autointerference, *Fed. Proc.*, 28, 1867, 1969.
11. **Sreevalsan, T.**, Homologous viral interference: Induction by RNA from defective particles of vesicular stomatitis virus, *Science*, 169, 991, 1970.
12. **Crick, J. and Brown, F.**, Interference as a measure of cross-relationship in the VSV group of rhabdoviruses, *J. Gen. Virol.*, 18, 79, 1973.
13. **Federer, K. B., Burrows, R., and Brooksby, J. B.**, Vesicular stomatitis virus — The relationship between some strains of the Indiana serotype, *Res. Vet. Sci.*, 8, 103, 1967.
14. **Cartwright, B. and Brown, F.**, Serological relationships between different strains of vesicular stomatitis virus, *J. Gen. Virol.*, 16, 391, 1972.
15. **Shope, R. E.**, Rabies virus antigenic relationships, in *The Natural History of Rabies*, Vol. 1, Baer, G. M., Ed., Academic Press, New York, 1975, 141.
16. **Dietzschold, B., Schneider, L. G., and Cox, J. H.**, Serology characterization of the three major proteins of vesicular stomatitis virus, *J. Virol.*, 14, 1, 1974.
17. **Cartwright, B., Talbot, P., and Brown, F.**, The proteins of biologically active subunits of vesicular stomatitis virus, *J. Gen. Virol.*, 7, 267, 1970.
18. **Cartwright, B., Smale, C. J., and Brown, F.**, Dissection of vesicular stomatitis virus into infective ribonucleoprotein and immunizing components, *J. Gen. Virol.*, 7, 19, 1970.
19. **Repik, P., Flamand, A., Clark, H F., Obijeski, J. F., Roy, P., and Bishop, D. H. L.**, Detection of homologous RNA sequences among six rhabdovirus genomes, *J. Virol.*, 13, 250, 1974.
20. **Bishop, D. H. L., Aaslestad, H. G., Clark, H F., Flamand, A., Obijeski, J. F., Repik, P., and Roy, P.**, Evidence for the sequence homology and genome size of rhabdovirus RNAs, in *Negative Strand Viruses*, Vol. 1, Mahy, B. W. J. and Barry, R. D., Eds., Academic Press, New York, 1975, 259.
21. **Sokol, F. and Koprowski, H.**, Structure-function relationships and mode of replication of animal rhabdoviruses, *Proc. Nat. Acad. Sci., U.S.A.*, 72, 933, 1975.
22. **Wagner, R. R.**, Reproduction of rhabdoviruses, in *Comprehensive Virology*, Vol. 4, Fraenkel-Conrat, H. and Wagner, R. R., Eds., Plenum Press, New York, 1975, 1.
23. **Wagner, R. R., Prevec, L., Brown, F., Summers, D. F., Sokol, F., and MacLeod, R.**, Classification of rhabdovirus proteins: A proposal, *J. Virol.*, 10, 1228, 1972.
24. **Bishop, D. H. L. and Roy, P.**, Dissociation of vesicular stomatitis virus and relation of the virion proteins to the viral transcriptase, *J. Virol.*, 10, 234, 1972.
25. **Kawai, A.**, Transcriptase activity associated with rabies virion, *J. Virol.*, 24, 826, 1977.
26. **Schneider, L. G. and Diringer, H.**, Structural and molecular biology of rabies virus, *Curr. Top. Microbiol. Immunol.*, 75, 153, 1976.
27. **Sokol, F.**, Chemical composition and structure of rabies virus, in *The Natural History of Rabies*, Baer, G. M., Ed., Academic Press, New York, 1975, 79.
28. **Wagner, R. R., Snyder, R. M., and Yamazaki, S.**, Proteins of vesicular stomatitis virus: Kinetics and cellular sites of synthesis, *J. Virol.*, 5, 548, 1970.
29. **Sokol, F., Stancek, D., and Kopowski, H.**, Structural proteins of rabies virus, *J. Virol.*, 7, 241, 1971.
30. **Emerson, S. V.**, Vesicular stomatitis virus: Structure and function of virion components, *Curr. Top. Microbiol. Immunol.*, 73, 1, 1976.
31. **Obijeski, J. F.**, unpublished observation, 1977.
32. **Wagner, R. R., Kiley, M. P., Snyder, R. M., and Schnaitman, C. A.**, Cytoplasmic compartmentalization of the proteins and ribonucleic acid species of vesicular stomatitis virus, *J. Virol.*, 9, 672, 1972.
33. **Soria, M. Little, S. P., and Huang, A. S.**, Characterization of vesicular stomatitis nucleocapsids. I. Complementary 40 S RNA molecules in nucleocapsids, *Virology*, 61, 270, 1974.
34. **Emerson, S. V. and Wagner, R. R.**, L protein requirement for *in vitro* RNA synthesis by vesicular stomatitis virus, *J. Virol.*, 12, 1325, 1973.
35. **Emerson, S. V. and Yu, Y. H.**, Both NS and L proteins are required for *in vitro* RNA synthesis by vesicular stomatitis virus, *J. Virol.*, 15, 1348, 1975.
36. **Szilágyi, J. F. and Uryvayev, L.**, Isolation of an infectious ribonucleoprotein from vesicular stomatitis virus containing an active RNA transcriptase, *J. Virol.*, 11, 279, 1973.
37. **Nakai, T. and Howatson, A. F.**, The fine structure of vesicular stomatitis virus, *Virology*, 35, 268, 1968.
38. **Murphy, F. A. and Harrison, A.**, Electron microscopy of rhabdoviruses, in *Rhabdoviruses*, Bishop, D. H. L., Ed., CRC Press, West Palm Beach, Fl., 1978, 4.

39. **Baltimore, D., Huang, A. S., and Stampfer, M.**, Ribonucleic acid synthesis of vesicular stomatitis. II. An RNA polymerase in the virion, *Proc. Nat. Acad. Sci., U.S.A.*, 66, 572, 1970.
40. **Rose, J. K. and Knipe, D.**, Nucleotide sequence complexities, molecular weights, and poly(A) content of the vesicular stomatitis virus in RNA species, *J. Virol.*, 15, 994, 1975.
41. **Baltimore, D.**, Expression of animal virus genomes, *Bacteriol. Rev.*, 35, 235, 1971.
42. **Bishop, D. H. L. and Roy, P.**, Properties of the product synthesized by vesicular stomatitis virus particles, *J. Mol. Biol.*, 58, 799, 1971.
43. **Bishop, D. H. L.**, Complete transcription by the transcriptase of vesicular stomatitis virus, *J. Virol.*, 7, 486, 1971.
44. **Coslett, G. D. and Obijeski, J. F.**, unpublished observation, 1977.
45. **Francki, R. I. B. and Randles, J. W.**, Composition of the plant rhabdovirus lettuce yellow necrotic virus in relation to its biological properties, in *Negative Strand Viruses*, Vol. 1, Barry, R. D. and Mahy, B. W. J., Eds., Academic Press, New York, 1975, 223.
46. **Wagner, R. R., Schnaitman, T. C., Snyder, R. M., and Schnaitman, C. A.**, Protein composition of the structural components of vesicular stomatitis virus, *J. Virol.*, 3, 611, 1969.
47. **Prevec, L.**, Physiological properties of vesicular stomatitis virus and some related rhabdoviruses, in *Viruses, Evolution and Cancer*, Kurstak, E. and Maramorosch, K., Eds., Academic Press, New York, 1974, 677.
48. **Kang, C. Y. and Prevec, L.**, Proteins of vesicular stomatitis virus. I. Polyacrylamide gel analysis of viral antigens, *J. Virol.*, 3, 403, 1969.
49. **Obijeski, J. F., Marchenko, A. T., Bishop, D. H. L., Cann, B. W., and Murphy, F. A.**, Comparative electrophoretic analysis of the virus proteins of four rhabdoviruses, *J. Gen. Virol.*, 22, 21, 1974.
50. **Wunner, W. and Pringle, C. R.**, Comparison of structural polypeptides from vesicular stomatitis virus (Indiana and New Jersey serotypes) and Cocal virus, *J. Gen. Virol.*, 16, 1, 1972.
51. **Freeman, G. J., Rose, J. K., Clinton, G. M., and Huang, A. S.**, RNA synthesis of vesicular stomatitis virus. VII. Complete separation of the mRNA's of vesicular stomatitis virus by duplex formation, *J. Virol.*, 21, 1094, 1977.
52. **Madore, H. P. and England, J. M.**, Rabies virus protein synthesis infected BHK-21 cells, *J. Virol.*, 22, 102, 1977.
53. **Stampfer, M. and Baltimore, D.**, Identification of the vesicular stomatitis virus large protein as a unique viral protein, *J. Virol.*, 11, 520, 1973.
54. **Knipe, D., Rose, J. K., and Lodish, H. F.**, Translation of individual species of vesicular stomatitis virus mRNA, *J. Virol.*, 15, 1004, 1975.
55. **Heyward, J. T. and Obijeski, J. F.**, unpublished observation, 1977.
56. **Sokol, F., Clark, H F., Wiktor, T. J., McFalls, M. L., Bishop, D. H. L., and Obijeski, J. F.**, Structural phosphoproteins associated with ten rhabdoviruses, *J. Gen. Virol.*, 24, 433, 1974.
57. **Kawai, A.**, Transcriptase activity associated wth rabies virion, *J. Virol.*, 24, 826, 1977.
58. **Knudson, D. L. and MacLeod, R.**, The proteins of potato yellow dwarf virus, *Virology*, 47, 285, 1972.
59. **Roy, P., Clark, H F., Madore, H. P., and Bishop, D. H. L.**, RNA polymerase associated with virions of pike fry rhabdovirus, *J. Virol.*, 15, 338, 1975.
60. **McAllister, P. E. and Wagner, R. R.**, Structural proteins of salmonid rhabdoviruses, *J. Virol.*, 15, 733, 1975.
61. **Clewley, J. P., Bishop, D. H. L., Kang, C. Y., Coffin, J., Schnitzlein, W. M., Reichmann, M. E., and Shope, R. E.**, Oligonucleotide fingerprints of RNA species obtained from rhabdoviruses belonging to the vesicular stomatitis virus subgroup, *J. Virol.*, 23, 152, 1977.
62. **Doel, T. R. and Brown, F.**, Tryptic peptide analysis of the structural proteins of vesicular stomatitis virus, *J. Gen. Virol.*, 38, 351, 1978.
63. **Moore, N. F. and Burke, D. C.**, Characterization of the structural proteins of different strains of Newcastle disease virus, *J. Gen. Virol.*, 25, 275, 1974.
64. **Laver, W. G. and Baker, N.**, Amino acid composition of polypeptides from influenza virus particles, *J. Gen. Virol.*, 27, 61, 1972.
65. **Kennedy, S. I. T. and Burke, D. C.**, Studies on the structural proteins of Semliki Forest virus, *J. Gen. Virol.*, 14, 87, 1972.

Chapter 8

STRUCTURE, SYNTHESIS, AND FUNCTION OF THE VESICULAR STOMATITIS VIRUS GLYCOPROTEIN

James R. Etchison and Donald F. Summers

TABLE OF CONTENTS

I. INTRODUCTION

The envelope of vesicular stomatitis virus (VSV) contains a single glycosylated protein species which has a molecular weight of approximately 67,000.[1] This glycoprotein constitutes the spike-like projections from the viral envelope which are seen in electron micrographs of negatively-stained, purified virus preparations.[2] Treatment of VSV with appropriate concentrations of a variety of proteases, such as trypsin, chymotrypsin, bromelain, thermolysin, or Pronase, selectively removes the glycoprotein from the virus particles leaving intact, subviral particles lacking the spike structures.[2,3] A fragment of the glycoprotein polypeptide chain remains associated with the lipid envelope of these subviral particles after trypsin treatment.[4] The glycoprotein may be selectively removed from virus particles in an intact form by treatment of purified virus preparations with nonionic detergents at low ionic strength.[5] This observation provides a basis for the purification of the glycoprotein that is rapid and that easily separates it from other virion components.[6-8]

II. STRUCTURE

Quantitative analysis of the purified glycoprotein has shown that it contains approximately 10% carbohydrate by weight.[7] This carbohydrate was found to consist of *N*-acetylneuraminic acid, galactose, *N*-acetylglucosamine, and mannose as the major constituents while fucose and *N*-acetylgalactosamine were detected in smaller amounts.[7,9] Extensive digestion of the glycoprotein with nonspecific proteases, such as Pronase®, yields oligosaccharide structures attached to a few residual amino acids.[10-12] These "glycopeptides" or "peptidyloligosaccharides" have been well characterized to determine the size, composition, and structure of the oligosaccharide moieties of the glycoprotein.[11] The peptidyloligosaccharides were found to contain an oligosaccharide moiety having an average molecular weight of approximately 3000 and an average of three or four amino acid residues.[11] Studies by Moyer et al.[12] showed that the oligosaccharides were attached to the peptide portion by a linkage having the alkaline stability characteristic of a 1-*N*-glycosylaminide bond of *N*-acetylglucosamine to the amide nitrogen of asparagine.[12]

The structure of the oligosaccharide moieties of the VSV glycoprotein has been elucidated by sequential chemical and enzymatic degradation.[11-13] The generalized structure for the oligosaccharide moieties is shown in Figure 1. This structure as shown is an idealized representation since the population of oligosaccharides is heterogeneous. A single structure within the population may contain either 0, 1, 2, or 3 residues of neuraminic acid (NeuNAc) and may or may not contain the single fucose (Fuc) residue.[11,14,15] This heterogeneity is depicted in Figure 1 by a "±" below the linkage. In addition to this "peripheral heterogeneity", other types of heterogeneity may also be present in a population of the oligosaccharides. An example of this is indicated in the structure in Figure 1 by the "±?" below one of the internal linkages. In this case all sugar residues distal to this linkage may be deleted in some of the structures present in the population. The linkages shown in the structure are for the oligosaccharides of the glycoprotein from VSV grown in BHK_{21} cells.[11,13] Differential sensitivity to an endoglycosidase preparation from *Streptococcus pneumoniae* has shown that the structures from virus grown in HeLa cells or polyoma-transformed BHK_{21} cells differs from those of virus grown in BHK_{21} cells.[12,16] The detailed aspects of these differences have not been defined.

There are two oligosaccharide chains of the structure shown in Figure 1 per polypeptide chain of the VSV glycoprotein.[7,11] Etchison and Holland calculated that, since the

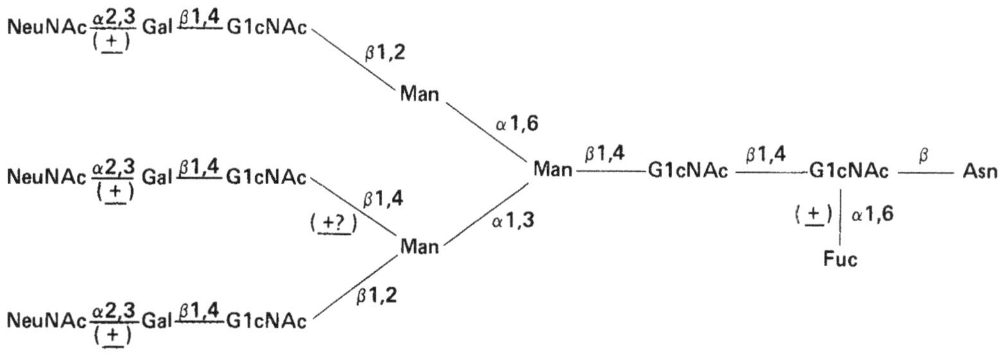

FIGURE 1.

glycoprotein contained approximately 10% carbohydrate by weight and since the molecular weight of the oligosaccharides was 3000 to 3400, there could be no more than an average of two such oligosaccharide chains per glycoprotein.[7] Robertson et al.[11] analyzed the tryptic peptides of the glycoprotein and showed that greater than 90% of the radiolabeled galactose, fucose, mannose, and *N*-acetylglucosamine incorporated into the glycoprotein was present in two of the peptides. The size of these two tryptic glycopeptides was analyzed and both were shown to be only a couple hundred daltons larger than the peptidyloligosaccharides obtained after Pronase digestion of the glycoprotein.[11] The tryptic glycopeptides were, therefore, not large enough to contain more than one oligosaccharide chain per tryptic peptide.

In addition to the above two oligosaccharide chains on the VSV glycoprotein, a third, minor oligosaccharide has been suspected. This is based on the detection of approximately one residue of *N*-acetylgalactosamine per glycoprotein.[7] A minor tryptic peptide species containing less than 5% of the glucosamine radiolabel was detected in the studies of Robertson et al.[11] Attempts to selectively label this minor peptide species with [^{14}C]galactosamine were unsuccessful since approximately 90% of the label present in the purified glycoprotein had been metabolized in the infected cells and was incorporated as *N*-acetylglucosamine rather than *N*-acetylgalactosamine (unpublished results). The presence or absence of such a minor oligosaccharide species remains to be unequivocally established.

The location or orientation of the two oligosaccharide moieties with respect to the polypeptide structure has not been determined. There are two basic questions involved in this aspect of the glycoprotein structure:

1. What is the location of the oligosaccharide chains in the primary structure of the polypeptide?
2. What is the orientation of the oligosaccharides in the three-dimensional, tertiary structure of the glycoprotein?

The first question is answerable by present sequencing technology and it should be possible to obtain many fragments of the polypeptide in relatively large amounts (several hundred micrograms). The second question is the more difficult to approach but, perhaps, the more important in terms of understanding the complete structure of the glycoprotein. Since the glycoprotein is insoluble unless detergent is present, it is unlikely that it could be crystallized. This makes it extremely unlikely that X-ray crystallography could be carried out on the intact glycoprotein. However, it should be possible to obtain a soluble fragment of the glycoprotein lacking only a small portion of the polypeptide and to crystallize this fragment for the X-ray studies. A similar approach has proven successful for the hemagglutinating glycoprotein of influenza virus and the X-ray crystallographic studies are underway.[18]

Indirect approaches may be conceived to address the two structural questions discussed above. These approaches, however, will necessarily involve assumptions which cannot be evaluated until the questions are answered directly. For example, a recent study allowed the temporal placement of the two glycosylation events with respect to the total time required to synthesize the polypeptide chain in an in vitro protein synthesizing and glycosylating system.[17] As discussed by the authors, this does not necessarily reflect the positional placement on the polypeptide chain. It is conceivable that the first glycosylation event may be further from the amino terminus of the polypeptide than is the second.

The orientation of the VSV glycoprotein in the viral envelope and in infected cell membranes has been studied by several approaches. The observation that the glycoprotein is the virion component which induces and binds neutralizing antibodies to VSV,[5,8] that the virion glycoprotein is sensitive to protease treatment,[2,3] and that the glycoprotein can be radiolabeled in vitro by reagents that do not penetrate the viral envelope[19,20] demonstrate that most of the glycoprotein structure is on the external surface of the virus particle. The lectin-binding studies of Penhoet et al.[21] indicate that the carbohydrate portion of the glycoprotein is also on the external surface of the viral envelope and infected-cell membranes. On the other hand, studies by Mudd and by Schloemer and Wagner have shown that not all of the polypeptide chain is located externally: a lipophilic fragment having a molecular weight of 5000 to 7000 remains imbedded in the lipid matrix of the envelope after protease treatment.[4,22]

Mudd and Swanson suggested that the lipophilic fragment is large enough to span the membrane and may, therefore, interact with other internal virion components.[23] Treatment of VSV virions with lipase or a mixture of the detergent Tween 80® and ether produced subviral particles from which lipid and the nonglycosylated membrane protein (M) had been removed.[24,25] Electron micrographs of these subviral particles suggested an interaction between the glycoprotein spikes and the ribonucleoprotein core. Recent studies by Mudd and Swanson demonstrated the presence of G and N heterodimers obtained after treatment of intact virions with reversible, protein cross-linking agents.[22] In addition, as was found by Dubovi and Wagner, heterodimers of G and M were also found.[26] These results suggest that the glycoprotein G may interact with either the M protein, the N protein, or both. That a highly specific interaction is not essential for virion stability and morphology has been demonstrated by the formation of VSV pseudotypes having the glycoproteins of myxoviruses, paramyxoviruses, or RNA tumor viruses during mixed infections with glycoprotein-defective mutants of VSV.[2,22,23,27-31]

The results of Schloemer and Wagner indicated that the lipophilic fragment associated wth the lipid envelope was derived from the amino terminus of the G polypeptide.[22] Katz et al.[32] suggested that the carboxy-terminal portion of the G protein spans the membrane. Studies which include sequencing both the amino- and carboxy-terminal regions of the G protein will be needed to clarify this apparent discrepancy. Studies to date have involved the assumption that either the amino-terminus or the carboxy-terminus is associated with the envelope. Figure 2 shows a schematic representation of the possible orientations of the polypeptide chain with respect to the asymmetry of the membrane or envelope Diagrams A and B show the two possible orientations considered above. Diagram C depicts the possibility that both the terminal sequences span the membrane, i.e., that both terminal sequences are protected from proteolytic digestion from without. The fourth possibility, shown in diagram D, predicts that both the

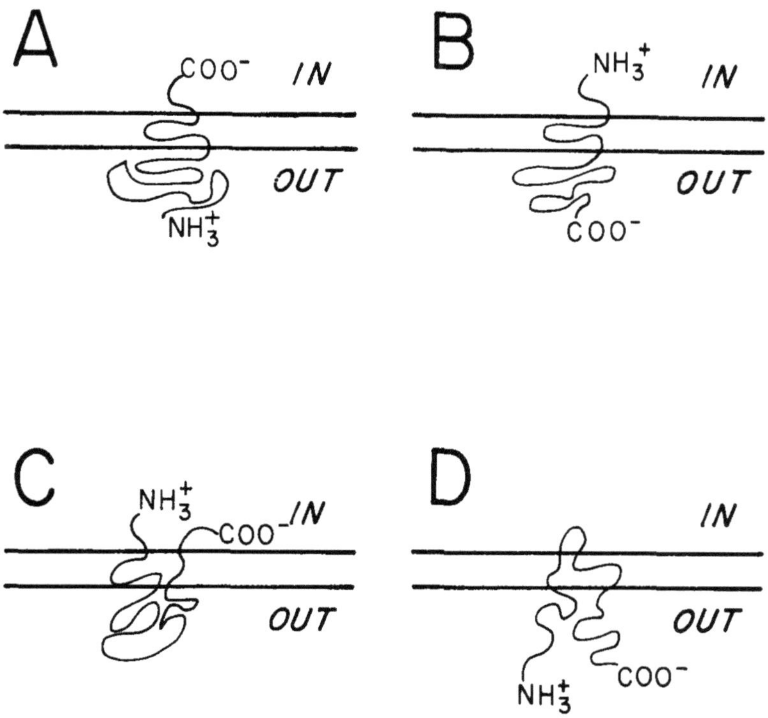

FIGURE 2.

terminal sequences would be degraded by protease digestion of the external surface. In the case shown in diagram C, the lipophilic "fragment" remaining after protease treatment of virions would be a mixture of both the terminal sequences. For the case in diagram D, the fragment would contain neither of the terminal sequences. Other possibilities, such as the location of a terminal sequence within the lipid matrix but not spanning the membrane, would not be distinguished from the possibilities shown by protease digestion techniques.

A higher order or quaternary structure for the G protein has not been established. In the protein crosslinking studies of Dubovi and Wagner and Mudd and Swanson both homodimers and homotrimers of the G protein were obtained.[23,36] Interpretation of these results is complicated by the possibility that, while the lipid matrix of the viral envelope is less fluid than that of the plasma membranes,[33] there may be lateral movement of proteins within the envelope. The finding of crosslinked dimers and trimers may be the result of fluidity rather than an indication that the spike structures of the viral envelope are dimers and/or trimers of the G protein.

III. SYNTHESIS

The mRNA coding for the VSV glycoprotein, in contrast to the other VSV specific mRNAs, is found exclusively on membrane bound polyribosomes in infected cells.[34-36] Newly synthesized G protein is found predominantly in the rough endoplasmic reticulum.[37,38] Recent studies suggest that the initiation of translation of G mRNA may occur on free ribosomes and a short portion of the peptide may be made prior to attachment of the mRNA-ribosome complex to the membranes in the rough endoplasmic reticulum.[39] Partial glycosylation of the polypeptide occurs during or immediately after translation in the rough endoplasmic reticulum.[38,40] Translation of the G mRNA in cell-free systems in the absence of membranes produces a nonglycosylated G polypeptide which has an apparent molecular weight of approximately 62,000.[36,41] The smallest G polypeptide observed after 5 or 10 min pulses in infected cells is approximately 65,000 daltons and is partially glycosylated.[38,40] Addition of membrane vesicle preparations to cell-free systems synthesizing G protein leads to the synthesis of a partially glycosylated species with the same molecular weight as the partially glycosylated species observed after short pulses in infected cells.[32,42,43]

The smallest size oligosaccharides attached to the VSV G protein detected in infected cells after short pulses with ^{3}H-glucosamine or ^{3}H-mannose were found to be at least as large as the core structure of the mature oligosaccharide structures (peptidyl-$GlcNAc_2Man_3$).[38] This result suggested that a preformed structure containing the oligomannosyl core structure may have been transferred *en bloc* to the nascent VSV G polypeptide. The donor of the preformed structure was presumed to be the polyisoprenylpyrophoshoryloligosaccharides found in various animal tissues and cultured cells.[44-46] The finding that tunicamycin, an antibiotic that blocks the formation of these polyisoprenyl-linked oligosaccharides, completely inhibited the glycosylation of the VSV G protein gave evidence for the involvement of these lipid-linked structures in the glycosylation of the VSV glycoprotein.[47]

The postulated role of these polyisoprenyl-linked oligosaccharides in the initial glycosylation of the VSV glycoprotein created an apparent enigma: The oligosaccharide structures attached to the polyisoprenyl structures contained several more mannose residues than those attached to the VSV glycoprotein.[48,49] This problem was answered by the findings of Hunt et al.[50] and Tabas et al.[51] that the initial glycosylation of the VSV glycoprotein did, in fact, involve the transfer of oversize oligomannose structures which were subsequently trimmed to mature-size oligomannose structures prior to a subsequent series of glycosylation reactions in which the distal sugar residues were added.

The trimming of the over-size oligomannose structures appears to involve at least two separable stages. Robertson et al.[52] and Tabas et al.[51] found that incomplete trimming occurs in lectin-resistant Chinese hamster ovary cell lines lacking a specific *N*-acetylglucosaminyltransferase. In the two different mutant cell lines studied, trimming appears to stop with structures containing five mannose residues in contrast to the processing to structures containing three mannose residues in the wild type or parental cell lines. These results suggest that the final stages of the trimming process may be coupled to the first of the final series of glycosylation reactions (i.e., the addition of the GlcNAc-Gal-NeuNAc branches) in an obligatory manner.

The total time required to complete the glycosylation and processing of the glycoprotein is not known. Atkinson et al.[53] have studied the kinetics of the VSV G maturation and found that there was a 20 min lag between the addition of fucose to the oligosaccharide moieties and the appearance of the mature G protein at the plasma membrane. There was no discernible lag between the appearance of the mature G at the plasma membrane and the incorporation of the mature G into progeny virus particles.

IV. FUNCTION

The VSV G protein forms the spikes or surface projections seen in electron micrographs of intact virions.[2,3] This peripheral location of the glycoprotein gives it a central role in several of the physiological properties of the virus particles.

The glycoprotein is the virion component which induces and binds neutralizing antibodies to VSV.[5,8,54] Kelley et al.[5] showed that the purified G protein induces and binds antibodies to VSV. Subsequently, Dietzschold et al.[8] showed that the virion glycoprotein is the only virion protein capable of inducing neutralizing antibodies to VSV; the M protein was inactive in this respect. It is not known whether the antigenic determinants of the glycoprotein reside in the polypeptide portion, the carbohydrate portion, or both. It is most likely the polypeptide portion that is antigenically active since Cartwright and Brown showed that a host specific antigen present in virus preparations was due to glycolipid and not the glycoprotein.[55]

The glycoprotein is involved in the adsorption (and penetration?) of virions to the host cell. Cartwright et al.[3] showed that treatent of virions with trypsin, which selectively removed the G protein, reduced infectivity by several logs. While studies have indicated involvement of the carbohydrate portion of the G protein in VSV adsorption, recent studies have shown that the oligosaccharide moieties play no essential role in the adsorption or penetration processes.[56-59] Schlesinger et al.[60] and Robertson et al.[52] grew VSV in lectin-resistant Chinese hamster ovary cells deficient in an *N*-acetylglucosaminyltransferase activity. The progeny virus particles contained G proteins lacking sialic acid and galactose and having only half as much *N*-acetylglucosamine as virion G from virus grown in the parental cell lines. These carbohydrate deficient virus particles had the same specific infectivity (plaque forming units per microgram protein of purified virus) as control preparations from parental or HeLa cells and similar yields of infectious progeny were obtained. Furthermore, Gibson et al.[61] have found that growth of VSV in the presence of tunicamycin, which blocks glycosylation of the VSV G protein, at 30° results in the production of reduced yields of virus particles (5 to 10% of controls) which contain G protein completely lacking carbohydrate. Nevertheless, these virus particles have the same specific infectivity as control, glycosylated G-containing virus particles.[61]

VSV and aggregates of the purified VSV glycoprotein are capable of hemagglutinating goose erythrocytes under stringently controlled conditions.[62,63] Variations in carbohydrate content of the glycoprotein may affect the efficiency of this hemagglutination reaction.[57,64] Since this reaction is specific for goose erythrocytes, it is probably not an important biological property of the VSV virion.

The VSV glycoprotein may be involved in aspects of the viral replication cycle other than the adsorption/penetration and maturation stages. Recent studies by McSharry and Choppin indicated that the purified G protein is able to cause an inhibition of cellular macromolecular synthesis.[65] Most temperature-sensitive mutations in the VSV G protein (complementation group V) do not appear to affect viral RNA synthesis and shutoff of host cell protein synthesis occurs at the nonpermissive temperature. However, Knipe et al.[66] recently reported that a complementation group V mutant of VSV was defective in both transcription and replication of viral RNA at the nonpermissive temperature. Since the VSV genome codes for the synthesis of only five proteins, the possibility that the virion "structural" proteins (G and M) play an intimate role in diverse areas of the viral replication process deserves further study.

REFERENCES

1. **Mudd, J. A. and Summers, D. F.,** Protein synthesis in vesicular stomatitis virus-infected HeLa cells, *Virology,* 42, 328, 1970.
2. **McSharry, J. J., Compans, R. W., and Choppin, P. W.,** Proteins of vesicular stomatitis virus and of phenotypically mixed vesicular stomatitis virus — simian virus 5 virions, *J. Virol.,* 8, 722, 1971.

3. **Cartwright, B., Smale, C. J., and Brown, F.,** Structural and biological relations in vesicular stomatitis virus, in *The Biology of Large RNA Viruses,* Barry, R. D. and Mahy, B. W. J., Eds., Academic Press, New York, 1970, 115.
4. **Mudd, J. A.,** Glycoprotein fragment associated with vesicular stomatitis virus after proteolytic digestion, *Virology,* 62, 573, 1974.
5. **Kelley, J. M., Emerson, S. U., and Wagner, R. R.,** The glycoprotein of vesicular stomatitis virus is the antigen that gives rise to and reacts with neutralizing antibody, *J Virol.,* 10, 1231, 1972.
6. **Bishop, D. H. L. and Roy, P.,**Dissociation of vesicular stomatitis virus and relation of the virion proteins to the viral transcriptase, *J. Virol.,* 10, 234, 1972.
7. **Etchison, J. R. and Holland, J. J.,** Carbohydrate composition of the membrane glycoprotein of vesicular stomatitis virus, *Virology,* 60, 217, 1974.
8. **Dietzschold, B., Schneider, L. G., and Cox, J. H.,** Serological characterization of the three major proteins of vesicular stomatitis virus, *J. Virol.,* 14, 1, 1974.
9. **McSharry, J. J. and Wagner, R. R.,** Carbohydrate composition of vesicular stomatitis virus, *J. Virol.,* 7, 412, 1971.
10. **Burge, B. W. and Huang, A. S.,** Comparison of membrane protein glycopeptides of Sindbis virus and vesicular stomatitis virus, *J. Virol.,* 6, 176, 1970.
11. **Etchison, J. R., Robertson, J. S., and Summers, D. F.,** Partial structural analysis of the oligosaccharide moieties of the vesicular stomatitis virus glycoprotein by sequential chemical and enzymatic degradation. *Virology,* 78, 375, 1977.
12. **Moyer, S. A., Tsang, J. M., Atkinson, P. H., and Summers, D. F.,** Oligosaccharide moieties of the glycoprotein of vesicular stomatitis virus, *J. Virol.,* 18, 167, 1976.
13. **Reading, C., Penhoet, E. E., and Ballou, C. E.,** Carbohydrate structure of vesicular stomatitis virus glycoprotein, *J. Supramol. Struct.,* (Suppl.1), 7, 1977.
14. **Robertson, J. S., Etchison, J. R., and Summers, D. F.,** Glycosylation sites of vesicular stomatitis virus glycoprotein, *J. Virol.,* 19, 871, 1976.
15. **Etchison J. R. and Holland, J. J.,** Carbohydrate composition of the membrane glycoprotein of vesicular stomatitis virus grown in four mammalian cell lines, *Proc. Nat. Acad. Sci. U.S.A.,* 71, 4011, 1974.
16. **Moyer, S. A. and Summers, D. F.,** Vesicular stomatitis virus envelope glycoprotein alterations induced by host cell transformation, *Cell,* 2, 63, 1974.
17. **Rothman, J. E. and Lodish, H. F.,** Synchronized transmembrane insertion and glycosylation of a nascent membrane protein, *Nature (London),* 269, 775, 1977.
18. **Brand, C. M. and Skehel, J. J.,** Crystalline antigen from the influenza envelope, *Nature (London) New Biol.,*238, 145, 1972.
19. **Walter, G. and Mudd, J. A.,** Iodination of vesicular stomatitis virus with lactoperoxidase, *Virology,* 52, 574, 1973.
20. **Moore, N. F., Kelley, J. M., and Wagner, R. R.,** Envelope proteins of vesicular stomatitis virus: accessibility to iodination, *Virology,* 61, 292, 1974.
21. **Penhoet, E., Olsen, C., Carlson, S., Lacorbiere, M., and Nicholson, G. L.,** Quantitative interaction of *Ricinus communis* agglutinin and Concanavalin A with influenza and vesicular stomatitis virus and virus-infected normal and polyoma-transformed cells, *Biochemistry,* 13, 3561, 1974.
22. **Schloemer, R. H. and Wagner, R. R.,** Association of vesicular stomatitis virus glycoprotein with membranes: isolation and characterization of a lipophilic fragment of the glycoprotein, *J. Virol.,* 16, 237, 1975.
23. **Mudd, J. A. and Swanson, R. E.,** *In situ* crosslinking of vesicular stomatitis virus proteins with reversible agents, *Virology,* 88, 263, 1978.
24. **Brown, F., Cartwright, B., Crick, J., and Smale, C.,** Infectivity of virus substructures from vesicular stomatitis virus, *J. Virol.,* 1, 368, 1967.
25. **Cartwright, B., Smale, C., and Brown, F.,** Surface structure of vesicular stomatitis virus, *J. Gen. Virol.,* 5, 1, 1969.
26. **Dubovi, E. J. and Wagner, R. R.,** Spatial relationships of the proteins of vesicular stomatitis virus: induction of reversible oligomers by cleavable protein cross-linkers and oxidation, *J. Virol.,* 22, 500, 1977.
27. **Choppin, P. W. and Compans, R. W.,** Phenotypic mixing of envelope proteins of parainfluenza virus SV-5 and vesicular stomatitis virus, *J. Virol.,* 5, 609, 1970.
28. **Zavada, J. and Rosenbergova, M.,** Phenotypic mixing of VSV with fowl plague virus, *Acta Virol. Engl. Ed.,* 16, 103, 1972.
29. **Love, D. N. and Weiss, R. A.,** Pseudotypes of vesicular stomatitis virus determined by exogenous and endogenous avian RNA tumor viruses, *Virology,* 57, 271, 1974.
30. **Zavada, J.,** VSV pseudotype particles with the coat of avian myeloblastosis virus, *Nature (London) New Biol.,*240, 122, 1972.

31. **Zavada, J.**, Pseudotypes of vesicular stomatitis virus with the coat of murine leukaemia and avian myeloblastosis virus, *J. Gen. Virol.*, 15, 183, 1972.
32. **Katz, F. N., Rothman, J. E., Lingappa, V. R., Blobel, G., and Lodish, H. F.**, Membrane assembly in vitro: synthesis, glycosylation, and asymmetric insertion of a transmembrane protein, *Proc. Nat. Acad. Sci. U.S.A.*, 74, 3278, 1977.
33. **Barenholz, Y., Moore, N. F., and Wagner, R. R.**, Enveloped viruses as model membrane systems: microviscosity of vesicular stomatitis virus and host cell membranes, *Biochemistry*, 15, 3563, 1976.
34. **Grubman, M. J., Moyer, S. A., Banerjee, A. K., and Ehrenfeld, E.**, Subcellular localization of vesicular stomatitis virus mRNA's, *Biochem. Biophys. Res. Commun.*, 62, 531, 1975.
35. **Grubman, M. J., Ehrenfeld, E., and Summers, D. F.**, *In vitro* synthesis of proteins by membrane-bound polyribosomes from vesicular stomatitis virus infected HeLa cells, *J. Virol.*, 14, 560, 1974.
36. **Both, G. W., Moyer, S. A., and Banerjee, A. K.**, Translation and identification of the mRNA species synthesized *in vitro* by the virion-associated RNA polymerase of vesicular stomatitis virus, *Proc. Nat. Acad. Sci. U.S.A.*, 72, 274, 1975.
37. **Hunt, L. A. and Summers, D. F.**, Association of vesicular stomatitis virus proteins with HeLa cell membranes and released virus, *J. Virol.*, 20, 637, 1976.
38. **Hunt, L. A. and Summers, D. F.**, Glycosylation of vesicular stomatitis virus glycoprotein in virus-infected HeLa cells, *J. Virol.*, 20, 646, 1976.
39. **Grubman, M. J., Weinstein, J. A., and Shafritz, D. A.**, Studies on the mechanism for entry of *vesicular stomatitis* virus glycoprotein G mRNA into membrane-bound polyribosome complexes, *J. Cell Biol.*, 74, 43, 1977.
40. **Knipe, D. M., Lodish, H. F., and Baltimore, D.**, Localization of two cellular forms of the vesicular stomatitis virus glycoprotein, *J. Virol.*, 21, 1121, 1977.
41. **Breindl, M. and Holland, J. J.**, Coupled *in vitro* transcription and translation of vesicular stomatitis virus messenger RNA, *Proc. Nat. Acad. Sci. U.S.A.*, 72, 2545, 1975.
42. **Toneguzzo, F. and Ghosh, H. P.**, Synthesis and glycosylation *in vitro* of glycoprotein of vesicular stomatitis virus, *Proc. Nat. Acad. Sci. U.S.A.*, 74, 1516, 1977.
43. **Toneguzzo, F. and Ghosh, H. P.**, *In vitro* synthesis of vesicular stomatitis virus membrane glycoprotein and insertion into membranes, *Proc. Nat. Acad. Sci. U.S.A.*, 75, 715, 1978.
44. **Waechter, C. J., Lucas, J. J., and Lennarz, W. J.**, Membrane glycoproteins. I. Enzymatic synthesis of mannosyl phosphoryl polyisoprenol and its role as a mannosyl donor in glycoprotein synthesis, *J. Biol. Chem.*, 248, 7570, 1973.
45. **Waechter, C. J. and Lennarz, W. J.**, The role of polyisoprenol-linked sugars in glycoprotein synthesis, *Annu. Rev. Biochem.*, 45, 95, 1976.
46. **Spiro, M. J., Spiro, R. G., and Bhoyroo, V. D.**, Lipid-saccharide intermediates in glycoprotein biosynthesis. I. Formation of an oligosaccharide-lipid by thyroid slices and evaluation of its role in protein glycosylation, *J. Biol. Chem.*, 251, 6400, 1976.
47. **Leavitt, R., Schlesinger, S., and Kornfeld, S.**, Tunicamycin inhibits glycosylation and multiplication of Sindbis and vesicular stomatitis viruses, *J. Virol.*, 21, 375, 1977.
48. **Spiro, R. G., Spiro, M. J., and Bhoyroo, V. D.**, Lipid-saccharide intermediates in glycopotein biosynthesis. II. Studies on the structure of an oligosaccharide lipid from thyroid, *J. Biol. Chem.*, 251, 6409, 1976.
49. **Robbins, P. W., Krag, S. S., and Liu, T.**, Effects of UDP-glucose addition on the synthesis of mannosyl lipid-linked oligosaccharides by cell-free fibroblast preparations, *J. Biol. Chem.*, 252, 1780, 1977.
50. **Hunt, L. A., Etchison, J. R., and Summers, D. F.**, Oligosaccharide chains are trimmed during synthesis of the envelope glycoprotein of vesicular stomatitis virus, *Proc. Nat. Acad. Sci. U.S.A.*, 75, 754, 1978.
51. **Tabas, I., Schlesinger, S., and Kornfeld, S.**, Processing of high mannose oligosaccharides to form complex type oligosaccharides on the newly synthesized polypeptides of the vesicular stomatitis virus G protein and the IgG heavy chain, *J. Biol. Chem.*, 253, 716, 1978.
52. **Robertson, M. A., Etchison, J. R., Robertson, J. S., Summers, D. F., and Stanley, P.**, Specific changes in the oligosaccharide moieties of VSV grown in different lectin-resistant CHO cells, *Cell*, 13, 515, 1978.
53. **Atkinson, P. H., Moyer, S. A., and Summers, D. F.**, Assembly of vesicular stomatitis virus glycoprotein and matrix protein into HeLa cell plasma membranes, *J. Mol. Biol.*, 102, 613, 1976.
54. **Kang, C. Y. and Prevec, L.**, Proteins of vescular stomatitis virus. II. Immunological comparisons of viral antigens, *J. Virol.*, 6, 20, 1970.
55. **Cartwright, B. and Brown, F.**, Glycolipid nature of the complement-fixing host cell antigen of vesicular stomatitis virus, *J. Gen. Virol.*, 15, 243, 1972.
56. **Schloemer, R. H. and Wagner, R. R.**, Sialoglycoprotein of vesicular stomatitis virus: role of the neuraminic acid in infection, *J. Virol.*, 14, 270, 1974.

57. **Schloemer, R. H. and Wagner, R. R.**, Cellular adsorption function of the sialo-glycoprotein of vesicular stomatitis virus and its neuraminic acid, *J. Virol.*, 15, 882, 1975.
58. **Schloemer, R. H. and Wagner, R. R.**, Mosquito cells infected with vesicular stomatitis virus yield virions of low infectivity, *J. Virol.*, 15, 1029, 1975.
59. **Cartwright, B. and Brown, F.**, Role of sialic acid in infection with vesicular stomatitis virus, *J. Gen. Virol.*, 35, 197, 1977.
60. **Schlesinger, S., Gottlieb, C., Feil, P., Gelb, N., and Kornfeld, S.**, Growth of enveloped RNA viruses in a line of Chinese hamster ovary cells with deficient N-acetylglucosaminyltransferase activity, *J. Virol.*, 17, 239, 1976.
61. **Gibson, R., Leavitt, R., Kornfeld, S., and Schlesinger, S.**, Synthesis and infectivity of vesicular stomatitis virus containing nonglycosylated G protein, *Cell*, 14, 671, 1978.
62. **Arstila, P., Halonen, P., and Salmi, G.**, Hemagglutinin of vesicular stomatitis virus, *Arch. Gesamte Virusforsch.*, 27, 198, 1969.
63. **Arstila, P.**, Small-sized hemagglutinin of vesicular stomatitis virus released spontaneously and with nonidet P-40, *Acta Pathol. Microbiol. Scand.*, 81, 27, 1973.
64. **McSharry, J. J., Ledda, C. A., Freiman, H. J., and Choppin, P. W.**, Biological properties of the VSV glycoprotein. II. Effects of host cell and of the glycoprotein carbohydrate composition on hemagglutination, *Virology*, 84, 183, 1978.
65. **McSharry, J. J. and Choppin, P. W.**, Biological properties of the VSV glycoprotein. I. Effects of the isolated glycoprotein on host macromolecular synthesis, *Virology*, 84, 172, 1978.
66. **Knipe, D., Lodish, H. F., and Baltimore, D.**, Analysis of the defects of temperature-sensitive mutants of vesicular stomatitis virus: intracellular degradation of specific viral proteins, *J. Virol.*, 21, 1140, 1977.

Chapter 9

VIRAL MEMBRANE PROTEIN STRUCTURE AND FUNCTION

James J. McSharry

TABLE OF CONTENTS

I. INTRODUCTION

Rhabdoviruses contain five structural proteins which have been designated L, G, N, NS, and M.[1] This chapter will deal with the structure and function of the M protein. This protein has been designated M for membrane because it is thought to be associated with the lipoprotein envelope of the virion. The evidence for the localization of the M protein with the viral envelope is as follows. Treatment with digitonin dissociates VSV into two fractions, a component consisting of membrane sheets containing proteins G and M and a nucleocapsid fraction containing proteins L, N, and NS.[2] These results suggest that M and G proteins are associated with the viral envelope. On the other hand, treatment of VSV with Triton X-100® solubilizes only the G protein, leaving the M protein in association with the nucleocapsid complex.[3,4] These results suggest that there is a strong interaction between the M protein and the nucleocapsid complex and may suggest that M is not a membrane-associated protein. However, both in vivo and in vitro studies suggest a clear association of the M protein with components of the viral envelope. Studies on the synthesis of VSV proteins showed that the M protein was synthesized on free polysomes and became rapidly associated with cellular membranes involved in virus maturation, whereas proteins of the nucleocapsid complex (L, N, and NS) were synthesized on free polysomes in the cytoplasm and did not associate with cellular membranes.[5-10] Furthermore, the M protein did not associate with the nucleocapsid complex while it was free in the cytoplasm.[11] Also in vitro experiments demonstrated that proteins M and G associated with cellular membranes whereas protein of the nucleocapsid complex, L, N, and NS did not associate with cellular membranes.[12] These results suggest that in the virion the M protein is associated with the viral envelope.

The membrane protein of VSV is thought to be associated with the internal side of the lipid bilayer. Evidence for this location of the M protein was obtained from the resistance of the M protein to protease digestion of intact virions and the failure to label the M protein with surface-labeling reagents in intact virions.[13-17] Recently it was shown that the M protein of VSV can be chemically cross-linked to both the nucleocapsid protein and the G protein suggesting that in the virion the M protein lies between these two proteins.[18] The accumulated data suggest that the M protein of VSV is associated with the viral envelope, but not exposed to the surface of the virion and lies between the proteins of the nucleocapsid complex and the surface glycoprotein.

Rabies virus contains two "membrane" proteins, M_1 and M_2. Treatment of rabies virus with deoxycholate resulted in a fraction of low buoyant density consisting of proteins M_1, M_2, and G and a nucleocapsid fraction containing protein N.[19] On the other hand, treatment of rabies virus with tri-(n-butyl) phosphate solubilized the G protein and the lipids from the envelope leaving the M_1 and M_2 proteins associated with the nucleocapsid complex.[20] These results are similar to the results of Triton X-100® treatment of VSV cited above and leave much doubt concerning the location and character of the two "membrane" proteins of rabies virus. Of particular interest would be the characterization of the M_1 and M_2 proteins with regard to their synthesis, location within the virion and association with the viral envelope or the nucleocapsid complex. The experiments described by Madore and England indicated that the synthesis of rabies virus proteins can be studied.[21] Extension of these studies to include cell fractionation in order to determine site of synthesis of the various viral proteins will help clarify the character of the M_1 and M_2 proteins. In addition, treatment of rabies virus with cross-linking reagents similar to those used to determine the location of VSV proteins should help determine the location of M_1 and M_2 proteins.[18]

Treatment of sowthistle yellow vein virus (SYVV) with Nonidet P-40® solubilized the lipid and proteins M_1, M_2, and G.[22] This result suggests that the M_1 and M_2 proteins of this plant rhabdovirus may be associated with the viral envelope. Although the synthesis and location of the M protein has been studied in detail for VSV Indiana only, all rhabdoviruses for which the polypeptide composition is known contain one or more "membrane" proteins with molecular weights similar to those described for VSV Indiana or rabies virus. Future studies on the synthesis of these proteins will help elucidate the true character of the M proteins from rhabdoviruses than other VSV Indiana.

II. CHEMICAL COMPOSITION

The VSV Indiana is the only rhabdovirus for which even the partial chemical composition of the membrane protein has been reported. Initially, polyethylene-glycol-dextran phase separation and isoelectric focusing techniques were used to isolate the viral membrane protein from VSV Indiana.[23,24] More recently, rhabdovirus proteins have been isolated by solubilizing the virions in 8*M* guanidine HCl followed by chromatography on Agarose A5M® or Sepharose B4® in 6*M* guanidine HCl.[17,25] Figure 1 shows the efficacy of this technique for isolating the VSV Indiana membrane protein. Column fractions 95 through 100 contain highly purified M protein. When guanidine is removed by dialysis the M protein precipitates and can be concentrated by low speed centrifugation. This procedure yields 1 to 2 mg of purified VSV membrane protein from approximately 10 mg of virus.

The hydrophobic character of the M protein has been suggested by its insolubility in aqueous solutions of Triton X-100® at low ionic strength and its solubility in guanidine HCl, sodium dodecyl sulfate, and Triton X-100 at high ionic strength.[3,17,25] The hydrophobic nature of the M protein was confirmed by amino acid analysis.[17,25] The amino acid composition of the membrane protein isolated by guanidine HCL column chromatography is shown in Table 1. The amino acid composition reveals the preponderance of neutral and hydrophobic residues, suggesting a very hydrophobic protein. Slight discrepancies between the amino acid composition reported in these two studies are probably due to differences in the preparation of the samples before amino acid analysis. In one study performic acid-oxidized protein was used for the determination of amino acid composition, whereas in the later study amino acid analysis was determined on untreated protein. The lack of cysteine in the one analysis is probably due to its destruction during hydrolysis in 6*N* HCL. According to the amino acid composition of the membrane protein, digestion with trypsin could lead to a maximum of 9 peptides. ^{35}S-methionine-labeled M protein was oxidized with performic acid and treated with TPCK-trypsin at 37°C for 24 hr.[26] The trypic peptides were separated by two-dimentional thin-layer chromatography.[27] Autoradiography revealed six ^{35}S-methionine-containing trypic peptides which is in close agreement to the number predicted from the amino acid composition.[28] The hydrophobic nature of the membrane protein correlates well with its presumed functions during virus maturation which include interactions with the hydrophobic tail of the VSV glycoprotein as well as with the hydrophobic nucleocapsid protein.[11,17,29,30]

III. FUNCTION

The function of the membrane protein has been studied using various serotypes of VSV. There is both direct and indirect evidence that the M protein is a group-specific antigen which is involved in maintaining virus structure, virus assembly, and possibly in regulation of viral RNA synthesis.

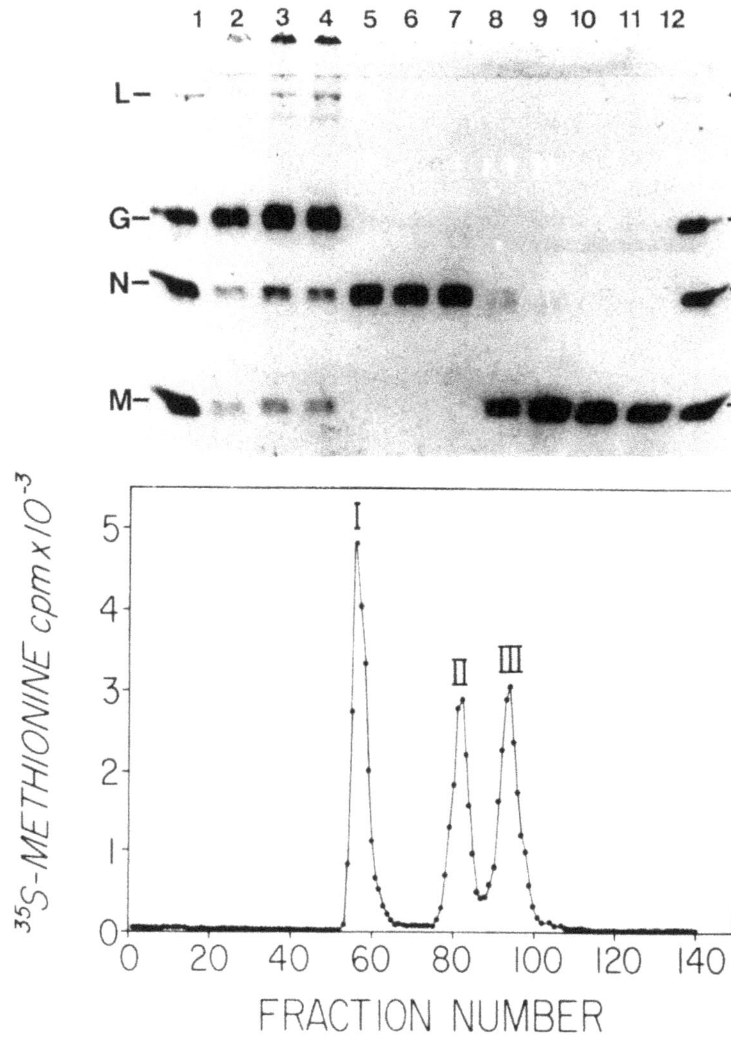

FIGURE 1. Agarose column chromatography of VSV proteins solubilized in guanidine HCl. The lower panel is the radioactive profile of ^{35}S-methionine labeled VSV effluent from the agarose column chromatography in 6*M* guanidine hydrochloride. The upper panel is a 10% polyacrylamide slab gel stained with Coomassie blue to identify the VSV polypeptides found in various fractions from the column. Slots on the extreme ends contain VSV polypeptides L, G, N/NS and M as markers. Slots 2, 3, and 4 contain fractions 55, 57 and 59 from peak I, slots 5, 6, and 7 contain fractions 81, 83 and 85 from peak II and slots 8, 9, 10 and 11 contain material from fractions 91, 93, 95 and 97 from column peak III. Results of PAGE show that peak II contains the nucleocapsid polypeptide N, and peak III contains VSV membrane protein M.

The isolated M protein is antigenic and reacts with mono-specific antibody in complement fixation and precipitation reactions.[24] The role of the M protein in the serology of rhabdoviruses has not been well investigated although there is indirect evidence that it is a group-specific antigen. The VSV nucleocapsid protein is a group-specific antigen and the nucleocapsid proteins of VSV Indiana and VSV Brazil are closely related antigenically whereas the nucelocapsid protein of VSV New Jersey is not antigenically related to the N protein of VSV Indiana or VSV Brazil.[31] Evidence that the M protein is a group-specific antigen with antigenic cross-reactions similar to those of the nucleo-

TABLE 1

Amino Acid Composition of the VSV Membrane Protein[11,25]

	Residues per mol	
Aspartate	20	26
Threonine	10	12
Serine	18	20
Glutamate	18	22
Proline	13	20
Glycine	16	19
Alanine	17	21
Valine	10	14
Methionine	6	11
Isoleucine	10	11
Leucine	15	21
Tyrosine	10	13
Phenylalanine	10	16
Histidine	8	9
Lysine	19	26
Arginine	9	10
Cysteine	2	—

capsid antigen have been recently reported.[32] The tryptic peptides of the membrane proteins isolated from the three strains of VSV have been compared. The proteins of VSV Brazil, VSV Indiana, and VSV New Jersey were iodinated, separated by SDS-polyacrylamide gel electrophoresis and the M protein of each virus strain was isolated from the gel. The iodinated M proteins were treated with trypsin and the peptides separated by electrophoresis and ascending chromatography on silica gel sheets. Autoradiograms revealed that the M proteins isolated from the three different serotypes of VSV shared most peptides suggesting that they are related. The relatedness of the peptides of the M proteins from these three serotypes was similar to that found for the peptides of the nucleocapsid protein from these three serotypes, i.e., the peptide maps of N and M proteins of VSV Indiana and VSV Brazil are closely related whereas neither are related to the peptide maps of N and M proteins of VSV New Jersey. Thus the M protein may also be a group-specific antigen which will be useful for classifying rhabdoviruses and comparing the relatedness between rhabdoviruses from plants, insects, fish, and mammals.

The membrane protein may be involved in maintaining the structure of VSV. Treatment of VSV with Tween-ether results in formation of VSV subviral particles which are infectious, fix complement, and retain their bullet-shape.[31,33] Subviral particles were also obtained by Nonidet-P40® treatment which removes both the G protein and the lipid. Treatment of this subviral particle with sodium deoxycholate destroys the bullet-shape structure of the subviral particle and separates it into a nucleocapsid component containing P3 and liberates P4, the membrane protein.[31] These results suggest that P4 helps to maintain the structure of the subviral particle. In addition, there is indirect evidence that the membrane protein may add stability to the viral envelope by lipid-protein interactions between the M protein and the phospholipids of the viral envelope.[34] Although most of the studies on the structure of the viral envelope have involved lipid-protein interactions between the external G protein and the viral lipids, only 10% of the lipid of the bilayer interacts with the G protein, suggesting that protein-lipid interactions which stabilize the lipid bilayer must involve the M protein.[35]

The membrane protein is thought to function in virus assembly. A specific interaction between the VSV nucleocapsid complex and the VSV membrane was demonstrated in phenotypically-mixed VSV SV5 hybrid viruses which contained VSV genetic information, all five VSV proteins, and the two SV5 glycoproteins.[15] The nucleocapsid and membrane proteins of SV5 are not found in the hybrid particles, suggesting that a specific interaction occurs between the VSV nucleocapsid complex and regions of the cell membrane which contained the VSV membrane protein. Interactions between the VSV nucleocapsid complex and patches of the plasma membrane which contained the SV5 membrane protein did not lead to the formation of bullet-shaped virions, although pleomorphic particles containing VSV nucleocapsids in an SV5 envelope could have been produced in these VSV- and SV5-infected cells. Recently a number of investigators have clarified the role of the VSV membrane in virus assembly. During VSV replication a soluble viral specific glycoprotein (Gs) is released into the medium. Using a temperature-sensitive mutant, *ts* G31, which produces temperature-sensitive M proteins, it was shown that at the nonpermissive temperature the synthesis of Gs is enhanced.[36,37] The results of these experiments suggest that the membrane protein functions to stabilize the attachment of the glycoprotein to the cell surface.The membrane protein also may be involved in forming a patch on the cell surface which contains a high concentration of G protein.[38] Treatment of VSV-infected cells with Concanavalin A (Con A) prevents virus release although the M and G proteins are synthesized and attached to the plasma membrane and the nucleocapsid (N) protein is free in the cytoplasm. Removal of the Con A with α-methyl D-glucoside results in release of infectious virus. These results can be interpreted to mean that the G proteins have to aggregate into a patch in order for virus release to occur. Specific interactions between the M protein and the hydrophobic tail of the G protein may be involved in aggregating the G proteins into the patch. In addition to its role in stabilizing the association of G proteins with the plasma membrane, the M protein facilitates the association of the nucleocapsid complex with patches of the plasma membrane which contain both M and G proteins.[10,11] This was demonstrated using temperature-sensitive mutants which had a defective M protein. At the nonpermissive temperature, M did not associate with the G protein, and the nucleocapsid complex did not attach to the plasma membrane of infected cells which contained normal G proteins. Thus the M protein has two roles in virus maturation: (1) to stabilize and aggregate the glycoproteins on the surface of infected cells, and (2) to allow the nucleocapsid to specifically associate with regions of the plasma membrane which contain viral-specific proteins. The interaction between the nucleocapsid and patches of the plasma membrane which contain viral-specific M and G proteins in the proper configuration initiates bud formation which results in maturation and release of virus from cell.

Recent evidence suggests that the M protein, possibly in cooperation with the G protein, may regulate virus transcription and replication.[39,40] When Group III *ts* mutants, which have a defective membrane protein, are grown at the nonpermissive temperature, an excess quantity of messenger RNA is synthesized, whereas at the permissive temperature normal amounts of RNA are synthesized. These results suggest that the M protein may inhibit the synthesis of messenger RNA. When either M or G proteins are nonfunctional, the synthesis of viral genomes is amplified, whereas under permissive conditions amplification occurs normally. These results suggest that these two envelope proteins may exert a positive control on VSV replication. Further studies on the role of envelope proteins in regulation of replication and transcription will be required in order to determine the role of M and/or G proteins in this process.

ACKNOWLEDGMENTS

I would like to thank the following investigators who contributed unpublished data for this review: Fred Brown, Alice Huang, Frank Landsberger, and Cristiane Printz.

REFERENCES

1. **Wagner, R. R., Prevec, L., Brown, F., Summers, D. F., Sokol, F., and McLeod, R.,** Classification of rhabdovirus proteins: a proposal, *J. Virol.,* 9, 672, 1972.
2. **Wagner, R. R., Schnaitman, T. C., Snyder, R. M., and Schnaitman, C. A.,** Protein composition of the structural components of vesicular stomatitis virus, *J. Virol.,* 3, 611, 1969.
3. **Emerson, S. U. and Wagner, R. R.,** Dissociation and reconstitution of the transcriptase and template activities of vesicular stomatitis B and T virions, *J. Virol.,* 10, 297, 1972.
4. **Kelley, J. M., Emerson, S. U., and Wagner, R. R.,** The glycoprotein of vesicular stomatitis virus is the antigen that gives rise to and reacts with neutralizing antibody, *J. Virol.,* 10, 1231, 1972.
5. **Wagner, R. R., Kiley, M. P., Snyder, R. M., and Schnaitman, C. A.,** Cytoplasmic compartmentalization of the proteins and ribonucleic acid species of vesicular stomatitis virus, *J. Virol.,* 9, 672, 1972.
6. **David, A. E.,** Assembly of vesicular stomatitis virus envelope: incorporation of viral polypeptides into the host plasma membrane, *J. Mol. Biol.,* 76, 135, 1973.
7. **Morrison, T. G. and Lodish, H. F.,** Site of synthesis of membrane and nonmembrane proteins of vesicular stomatitis virus, *J. Biol. Chem.,* 250, 6955, 1975.
8. **Atkinson, P. H., Moyer, S. A., and Summers, D. F.,** Assembly of vesicular stomatitis virus glycoprotein and matrix protein into HeLa cell plasma membranes, *J. Mol. Biol.,* 102, 613, 1976.
9. **David, A. E.,** Assembly of the vesicular stomatitis virus envelope: transfer of viral polypeptides from polysomes to cellular membranes, *Virology,* 76, 98, 1977.
10. **Knipe, D. P., Baltimore, C., and Lodish, H. F.,** Separate pathways of maturation of the major structural proteins of vesicular stomatitis virus, *J. Virol.,* 21, 1128, 1977.
11. **Knipe, D. M., Baltimore, D., and Lodish, H. F.,** Maturation of viral proteins in cells infected with temperature-sensitive mutants of vesicular stomatitis virus, *J. Virol.,* 21, 1149, 1977.
12. **Morrison, T. and McQuain, C.,** Assembly of viral membranes. I. Association of vesicular stomatitis virus membrane proteins and membranes in a cell free system, *J. Virol.,* 21, 451, 1977.
13. **Cartwright, B., Small, C. J., and Brown, F.,** Surface structure of vesicular stomatitis virus, *J. Gen. Virol.,* 5, 1, 1967.
14. **Cartwright, B., Talbot, P., and Brown, F.,** The proteins of biologically active subunits of vsicular stomatitis virus, *J. Gen. Virol.,* 7, 267, 1970.
15. **McSharry, J. J., Compans, R. W., and Choppin, P. W.,** Proteins of vesicular stomatitis virus and of phenotypically mixed vesicular stomatitis virus-simian virus 5 virions, *J. Virol.,* 8, 722, 1971.
16. **Eger, R., Compans, R. W., and Rifkin, D. B.,** The organization of the proteins of vesicular stomatitis virions: labeling with pyridoxal phosphate, *Virology,* 66, 610, 1975.
17. **McSharry, J. J.,** The effect of chemical and physical treatments on the lactoperoxidase-catalyzed iodination of vesicular stomatitis virus, *Virology,* 83, 482, 1977.
18. **Dubovi, E. J. and Wagner, R. R.,** Spatial relationships of the proteins of vesicular stomatitis virus: induction of reversible oligomers by cleavable protein cross-linkers and oxidation, *J. Virol.,* 22, 500, 1977.
19. **Sokol, F., Stancek, D., and Koprowski, H.,** Structural proteins of rabies viruses, *J. Virol.,* 7, 241, 1971.
20. **Neurath, A. R., Vernon, S. K., Dobkin, M. B., and Rubin, B. A.,** Characterization of subviral components from treatment of rabies virus with tri-(n-butyl) phosphate, *J. Gen. Virol.,* 14, 33, 1972.
21. **Madore, H. P. and England, H. M.,** Rabies virus protein synthesis in infected BHK-21 cells, *J. Virol.,* 22, 102, 1977.
22. **Ziemiecki, A. and Peters, D.,** The proteins of sowthistle yellow vein virus: characterization and location, *J. Gen. Virol.,* 32, 369, 1976.
23. **Bishop, D. H. L. and Roy, P.,** Dissociation of vesicular stomatitis virus and relation of the virion proteins to the viral transcriptase, *J. Virol.,* 10, 234, 1972.
24. **Dietzschold, B., Schneider, L. G., and Cox, H. J.,** Serological characterization of the three major proteins of vesicular stomatitis virus, *J. Virol.,* 14, 1, 1974.
25. **Kendal, A. P. and Cohen, P.,** Isolation of the matrix (membrane) protein of vesicular stomatitis virus by gel filtration in guanidine hydrochloride, *J. Gen. Virol.,* 33, 543, 1976.
26. **Hirs, C. H. W.,** Determination of cysteine as cysteic acid, in *Methods in Enzymology,* Vol. 11, Hirs, C. H. W., Ed., Academic Press, New York, 1967, 59.
27. **Bellisario, R., Carlsen, R. B., and Bahl, O. P.,** Human chorionic gonadotropin: linear amino acid sequence of the α-subunit, *J. Biol. Chem.,* 248, 6796, 1973.
28. **McSharry, J. J.,** Unpublished data, 1977.
29. **Mudd, J. A.,** Glycoprotein fragment associated with vesicular stomatitis virus after proteolytic digestion, *Virology,* 62, 573, 1974.

30. **Schloemer, R. H. and Wagner, R. R.,** Association of vesicular stomatitis virus glycoprotein with virion membrane: characterization of the lipophilic tail fragment, *J. Virol.,* 16, 237, 1975.
31. **Cartwright, B. and Brown, F.,** Serological relationships between different strains of vesicular stomatitis virus, *J. Gen. Virol.,* 16, 391, 1972.
32. **Doel, T. R. and Brown, F.,** Tryptic peptide analysis of the structural proteins of vesicular stomatitis virus, *J. Gen. Virol.,* in press 38, 351, 1978.
33. **Brown, F., Cartwright, B., Crick, J., and Smale, C. J.,** Infective substructure from vesicular stomatitis virus, *J. Virol.,* 1, 368, 1967.
34. **Stoffel, W., Bister, K., Schreiber, C., and Tunggal, B.,** ^{13}C-NMR studies of the membrane structure of enveloped viruses (vesicular stomatitis virus), *Z. Physiol. Chem.,* 357, 905, 1976.
35. **Landsberger, F.,** Personal communication, 1978.
36. **Little, S. P. and Huang, A. S.,** Synthesis and distribution of vesicular stomatitis virus-specific polypeptides in the absence of progeny production, *Virology,* 81, 37, 1977.
37. **Little, S. P. and Huang, A. S.,** Shedding of the glycoprotein from vesicular stomatitis virus-infected cells, *J. Virol.,* 27, 330, 1978.
38. **Cartwright, B.,** Effect of concanavalin A on vesicular stomatitis virus maturation, *J. Gen. Virol.,* 34, 249, 1977.
39. **Printz, C.,** Personal communication, 1977.
40. **Huang, A. S., Little, S. P., and Clinton, G.,** Personal communication, 1977.

Chapter 10

VIRAL TRANSCRIPTASE

D. M. Hunt, M. G. Mellon, and S. U. Emerson

TABLE OF CONTENTS

I. INTRODUCTION

Until the late 1960's it was assumed that virus particles were too simple to contain enzymes. However, in 1967 it was shown that vaccinia virions carried a DNA-dependent RNA polymerase.[1,2] This observation was soon followed by reports of an RNA-dependent RNA polymerase in virions of double stranded RNA viruses.[3-5] Then in 1970, Baltimore, Huang, and Stampfer determined that virions of the negative-stranded rhabdovirus vesicular stomatitis virus (VSV) also contained an RNA-dependent RNA polymerase.[6] Because rhabdoviruses are relatively simple viruses which are easily purified, this demonstration initiated a multitude of studies on rhabdovirus RNA synthesis. The VS virion polymerase, or transcriptase, has since been shown to transcribe the viral genome into 5 monocistronic mRNAs which are capped, polyadenylated, and methylated;[7] in addition, cleavage of the initial transcript has been postulated.[7] Therefore, this supposedly simple virus particle carries the enzyme or enzymes required to catalyze many complex biochemical reactions.

Following the discovery of a polymerase in VSV, other rhabdoviruses were also shown to contain a similar enzyme activity. Purified virions of the animal rhabdoviruses, VSV New Jersey, VSV Indiana, Cocal, Chandipura, Piry, and Kern Canyon all contain a transcriptase.[8,9] Fish rhabdoviruses such as pike fry virus, infectious hematopoietic necrosis virus and virus of hemorrhagic septicemia similarly package a transcriptase in their virions.[10,11] A transcriptase has also been discovered in virions of the plant rhabdovirus, lettuce necrotic yellows virus.[12] Although the general properties of these viral enzymes are similar, the specific activities differ drastically.[9] Attempts to demonstrate a transcriptase in rabies virions were initially unsuccessful leading to speculation that rabies virus might be unique. However, recently it was reported that the conditions for rabies transcription in vitro were slightly different than those of the other rhabdoviruses.[13] If a relatively high concentration of nonionic detergent is used to activate the rabies virions, a transcriptase activity can be demonstrated in this rhabdovirus also. Since transcriptases have now been described in animal, fish, and plant rhabdoviruses it appears most reasonable to assume that the presence of a virion associated transcriptase is a characteristic of all rhabdoviruses.

All of the rhabdovirus transcriptases thus far examined display in vitro requirements for activity similar to those originally observed by Baltimore et al.[6] for VSV Indiana (Table 1). If intact virions are assayed, a nonionic detergent is required to disrupt the virions. All four nucleoside triphosphates are required for RNA synthesis and a reducing agent such as dithiothreitol or mercaptoethanol is needed for maximum transcription. The optimum salt concentration of the reaction mixture varies somewhat with the particular virus but, in general, concentrations ranging from 0.1 to 0.15 *M* NaCl are used. In addition, Mg^{++} is required and cannot be replaced by Mn^{++} except in the case of the salmonid viruses which can use either. The temperature optimum does differ according to the class of viruses studied. The animal rhabdoviruses have a temperature optimum around 30°C while the fish rhabdoviruses which normally grow in colder hosts consequently have a much lower temperature optimum between 15 to 20°C.

Rhabdovirus virion RNA is not infectious. This fact led Baltimore et al.[6] to suggest that rhabdoviruses might be obliged to carry a polymerase within the virion in order to initiate the infectious cycle. VSV grows in enucleated cells or actinomycin D treated cells indicating that no short lived host proteins participate in the viral infection.[14] In addition, the tremendous host range of VSV suggests that host cell proteins either do not play a central role in VS virus replication or that the host proteins involved are widely distributed. In addition, considerable evidence now indicates that the virion

TABLE 1

In Vitro Transcription Conditions for Various Rhabdoviruses

Virus	Detergent	Salt	Metal ion	Temperature	Ref.
Vesicular stomatitis virus	0.08% Triton N101®	100 m*M* NaCl	Mg^{++}	37°C[a]	6
Chandipura, Piry, Cocal vesicular stomatitis virus	1% Triton N101®	104 m*M* NaCl	Mg^{++}	31°C	9
Kern Canyon virus	1% Triton N101®	104 m*M* NaCl	Mg^{++}	31°C	8
Infectious hematopoietic necrosis virus	0.37% Triton X 100®	144 m*M* NaCl	Mg^{++} or Mn^{++}	18°C	11
Virus of hemorrhagic septicemia	0.37% Triton X 100®	144 m*M* NaCl	Mg^{++} or Mn^{++}	15°C	11
Pike fry virus	1% Triton N101®	104 m*M* NaCl	Mg^{++}	20°C	10
Rabies virus	4—5% Triton × 100®	40 m*M* NaCl	Mg^{++}	30°C	13
Lettuce necrotic yellows virus	1% Nonidet P-40®	125 m*M* NaCl	Mg^{++}	25°C	12

[a] The original experiments were done at 37°C; subsequently it was shown that the optimum temperature for in vitro transcription of VSV was closer to 31°C.

transcriptase is composed mainly, if not totally, of virally coded proteins and functions in vivo to synthesize viral mRNA. Viral mRNAs isolated from infected cells or synthesized in vitro are qualitatively similar.[7] The in vitro translation products of these RNAs are also basically the same, thus establishing the fidelity of the in vitro transcription.[15] Finally genetic analyses indicate that viral proteins are required for transcription. VSV Indiana or Cocal temperature sensitive (*ts*) mutants can be assigned to five, or possibly six complementation groups.[16] Three of the classes have mutants which display decreased RNA synthesis at the nonpermissive temperature. Similarly, VSV New Jersey *ts* mutants can be divided into six complementation groups, three of which are defective in RNA synthesis; a fourth class contains some mutants which do not synthesize RNA at nonpermissive temperature while others do. In summary, the genetic studies imply that at least three viral proteins directly affect VSV RNA synthesis. As discussed below, biochemical studies support this conclusion.

VSV is the prototype of the rhabdoviruses and has been studied most extensively. Therefore, the remainder of this chapter will focus mainly on the function and structure of the VSV enzyme with the underlying assumption that similar characteristics will eventually be demonstrated for the RNA-dependent RNA polymerases of other rhabdoviruses.

II. SUBPARTICLES OF VSV CONTAIN TRANSCRIPTASE

Purified VS virions can be stripped of their lipid envelope and the membrane proteins G and M with neutral detergent and moderate ionic strength solutions. The insoluble subparticles or ribonucleocapsid cores which result from this treatment are capable of transcription.[17-19] The cores contain the major viral protein N tightly associated with virion RNA and the minor proteins L and NS. Since the transcriptase activity of the virion does not decrease following the removal of G and M proteins, neither of these proteins is apparently required for transcription. The RNA products synthesized in vitro by the viral cores are identical to VSV messenger RNA species and have been translated into serotype specific VSV proteins.[20,21] The nucleocapsid cores, which appear as helical, filamentous structures in the electron microscope,[18] are infectious.[19]

The nucleocapsid cores are similar to ribonucleoprotein complexes isolated from infected cells. These complexes synthesize polyadenylated RNA complementary to virtually all regions of the genome.[22,23] Like the subviral cores, the transcribing complexes isolated from infected cells contain N and L proteins and 42S RNA.[24] Although not detected in this study, the NS protein is presumably also present in this preparation (see below).

III. IN VITRO RECONSTITUTION OF THE VSV TRANSCRIPTASE

Emerson and Wagner dissociated and reconstituted the transcriptase complex of VSV.[18] This reconstitution assay has proved a powerful tool for analyzing the structural requirements for transcription. Treatment of virions with a high salt (0.72 *M* NaCl)-Triton containing buffer followed by high speed centrifugation removes not only the G and M proteins but also the L and NS proteins from the insoluble nucleocapsid core. Neither the solubilized supernatant fraction nor the insoluble core was capable of transcription alone, but substantial activity was restored on their recombination. Since the G and M proteins are not necessary for transcription, it was concluded that either the L or NS protein or both were required. While the experiments

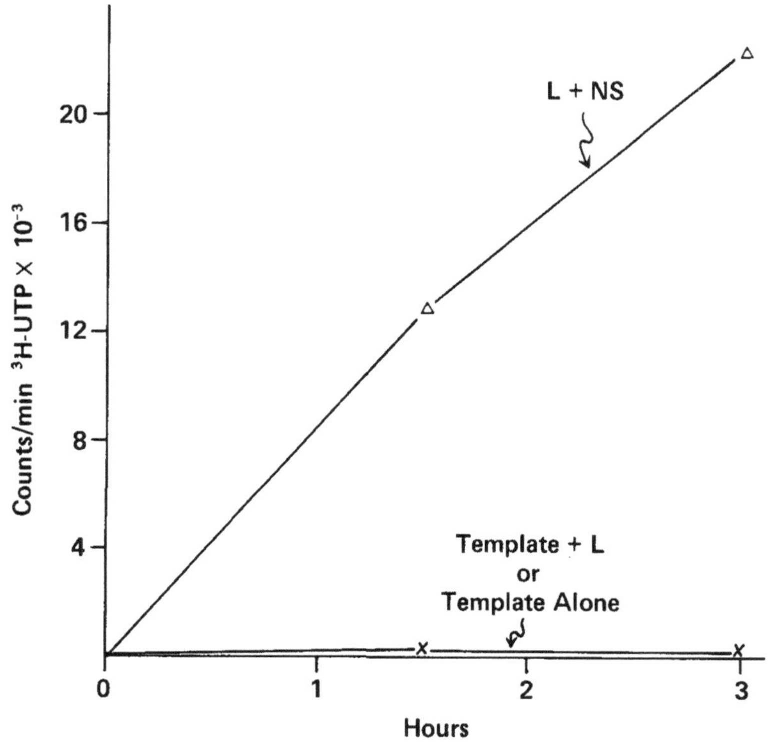

FIGURE 1. Reconstitution experiment demonstrating that both L and NS proteins of VSV-Indiana are required for in vitro transcription. Purified VSV-Indiana template containing N protein and viral RNA was assayed alone or in combination with L or L and NS protein in a standard transcription assay. Symbols: template alone (x), template and L protein (x), template and both L and NS proteins (Δ). (From Emerson, S. U. and Yu, Y-H., *J. Virol.*, 15, 1348, 1975. With permission.)

described above were done with VSV-Indiana virus, subsequently the same methods were used to dissociate and reconstitute an active transcriptase in three other members of the VS virus subgroup — VSV New Jersey, VSV Cocal, and Chandipura.[25] Additionally, in each case the reconstituted subparticles were infectious. The RNA products synthesized by the reconstituted transcriptase are complementary to the virion RNA template, similar in size to VSV messages, and like authentic messages are capped, methylated, and polyadenylated.[20]

Emerson and Yu further purified and separated the L and NS proteins and added them individually to highly purified template fraction which contained only N protein bound to virion RNA. These experiments demonstrated that both the L and NS proteins are required for transcription (Figure 1).[26,27] Naito and Ishihama have confirmed the requirement for NS protein.[28] They have purified an NS:L protein complex, free of M and G proteins. Fractions of the purified NS:L protein complex which contain less than a 1:1 stoichiometric ratio of NS:L protein are proportionately less active in reconstitution transcriptase assays than complexes containing molar equivalents of both proteins. Inclusion of additional NS protein stimulated transcriptase activity in these mixtures.

Deproteinized viral RNA (as well as various other natural and synthetic nucleic acids) will not serve as template in the reconstituted system.[18,28,29] Thus only virion RNA coated with N protein will serve as template and all three viral proteins, N, NS, and L are required for transcription.

IV. THE SPECIFICITY OF THE TRANSCRIPTASE WITHIN THE VSV SUBGROUP OF VIRUSES

Using in vitro reconstitution assays Bishop et al. examined the stringency of the transcriptase reaction by analyzing heterologous combinations of supernatant and template fractions prepared from several members of the VSV subgroup of viruses.[25] The transcriptase fractions of VSV Indiana and VSV Cocal are active when recombined with the template of either virus. However, no significant levels of trancription were detected in combinations of VSV Indiana or VSV Cocal supernatant fractions with templates from either VSV New Jersey or Chandipura viruses. Reciprocal combinations using New Jersey or Chandipura supernatants and VSV Indiana or VSV Cocal templates were also inactive. Thus with the exception of the closely related VSV Cocal and VSV Indiana viruses, the transcriptase enzyme appears to function only with its own template. Additional experiments were done by Emerson and Yu to determine whether the L or NS protein or both were responsible for the serotype specificity of the transcriptase.[30] In these experiments three fractions, a template fraction and two supernatant fractions containing either the L or NS protein, were prepared from both VS Indiana and VS New Jersey viruses. The L and NS protein fractions were mixed in vitro in homologous or heterologous combinations with either the VSV New Jersey or VSV Indiana template fractions. The only combinations which were capable of detectable transcription were those in which all three fractions were obtained from a homologous serotype virus. Substitution of either L or NS protein fraction from the heterologous serotype virus prohibited transcription. Thus the serotype specificity observed for the transcriptase in unfractionated supernatants extends to both the L and NS proteins.

The in vitro results contrast with a report by Repik et al. of the in vivo complementation of a mutant VSV Indiana by wild type VSV New Jersey.[31] *Ts* G-I-114 is a Group I mutant of Indiana serotype which is temperature sensitive for primary (cycloheximide-resistant) RNA transcription in the infected cell. Coinfection of G-I-114 with wild type New Jersey at nonpermissive temperatures allows synthesis of RNA complementary to the Indiana genome. The presence of puromycin prevents the complementation suggesting that some *de novo* gene product of New Jersey virus can aid in the transcription of the VSV Indiana mutant. Since the *ts*-G-I-114 lesion is now known to be in the L gene,[32] it seems that some protein product of New Jersey virus can stabilize or replace the defective L gene in the mutant in vivo.

V. THE TRANSCRIPTASE OF DEFECTIVE PARTICLES OF VSV

Populations of VSV often include one or more types of particles, shorter in length than the wild type virion, which reduce the yield of wild type virus in mixed infections.[33] These defective interfering (DI) particles contain L and NS proteins in about the same amounts relative to other viral proteins as wild type virions.[34,35] All defective particles isolated so far are derived from the L region or the 5′ end of the standard genome, with the exception of one defective particle which is derived from the 3′ end of the standard genome and carries the genes for N, NS, M, and G proteins.[36-38] DI particles derived from both ends of the genome have been shown to contain an active transcriptase.

The purified defective particle from the 3′ end can transcribe in vitro the 4 genes carried on its abbreviated genome as well as a small "leader" sequence of RNA and thus clearly has an active transcriptase.[38] The demonstration that defective particles derived from the L region of the genome also carry an active transcriptase is more complicated. One of these defective particles has been shown to synthesize a unique, 45 nucleotide piece of RNA, but like the other L region DI particles it does not transcribe messages in vitro.[39,40] However, mixing experiments show that the transcriptase carried by this DI particle is capable of transcribing more than the small 45 nucleotide long product. If combined with a wild type template, the supernatant fraction solubilized from the L region defective particle will support high levels of transcription, comparable to the wild type supernatant.[18] The reciprocal observation that the defective particle template fractions are only marginally active even when tested with a wild type supernatant fraction suggests that while the transcriptase enzyme of the L region DI particle is functional, the template somehow restricts transcription to only the small 45 nucleotide species of RNA.

VI. ROLES OF THE 3 TRANSCRIPTION PROTEINS

The precise roles of the three viral proteins required for transcription are not yet known. As the major structural protein of the virus, the N protein is almost certainly not the transcriptase enzyme, but until transcription can be obtained on a template in the absence of N protein, the possibility cannot be rigorously excluded. The basis for the N protein requirement may be that it constitutes part of the recognition site for the enzyme or that it confers a special configuration on the viral RNA. The L and NS proteins are the probable candidates for the transcriptase enzyme. The L protein is a large, unique polypeptide of 160 to 190,000 daltons.[27,33,41] Because of its large size, similar to certain *E. coli* transcriptases,[42] and its ability to bind to phosphocellulose, the L protein probably functions as the actual catalytic transcriptase subunit. On the other hand, no specific role has as yet been proposed for the NS protein. NS protein is a heat stable phosphoprotein,[26,43] whose molecular weight has not yet been satisfactorily established. As determined on standard Maizel SDS gels, the molecular weight of NS protein is 40,000 to 45,000 daltons.[33] However, the coding capacity of the NS mRNA is sufficient for a protein no larger than 30,000 daltons.[44-46] Naito and Ishihama suggest that NS protein is an intrinsic component of the transcriptase enzyme since it can be isolated in intimate association with L protein.[28] Antibody to the NS protein will immediately terminate transcription when added to an in vitro system.[47]

Mellon and Emerson have developed an assay, based on the reconstitution of the VSV transcriptase, for the binding of the L and NS proteins to the N-coated-RNA template.[48] When the high salt solubilized supernatant fraction of VS New Jersey virus is recombined under transcriptase conditions with a purified template fraction, only the L and NS proteins bind to template (Figure 2). This is true even though the M and G proteins are the most abundant proteins in the supernatant. The template can be saturated with L and NS proteins. At saturation the numbers of L and NS proteins bound per 1000 N molecules are equivalent to the numbers of L and NS molecules per 1000 N molecules in the intact virus, suggesting that the binding measured in the assay mimics the in vivo association of transcriptase and template. Purified NS protein binds to the template in the absence of L protein. However, the L protein apparently does not bind to template in the absence of NS protein. Several explanations are possible for this observation. The NS protein may simply stabilize the quite labile L protein. Alternatively, the NS protein may serve to anchor the L protein to the template or it may modify the template in some way as to expose L protein binding sites.

VII. IN VITRO STUDIES OF TRANSCRIPTION PROTEINS

Study of mutants can offer further insights into the function of virion proteins and

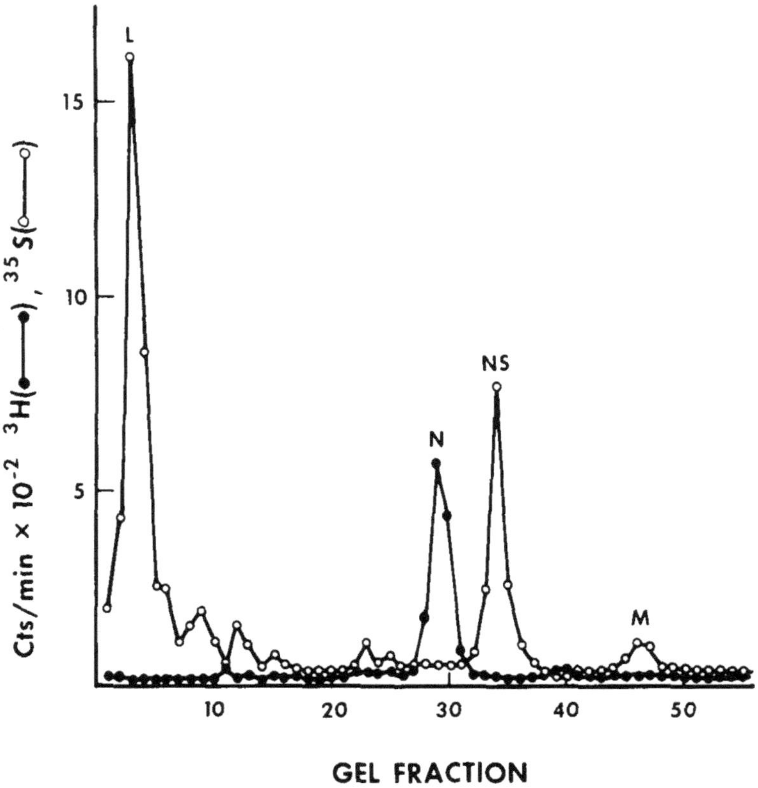

FIGURE 2. The rebinding of L and NS proteins to purified nucleocapsid. ^{35}S labeled virus was solubilized with a high salt-triton solution and the nucleocapsids were removed by centrifugation. The solubilized ^{35}S labeled M, G, L, and NS proteins were recombined under transcription conditions with a ^{3}H-labeled template containing only N protein and viral RNA. The nucleocapsids were separated from unbound proteins on a glycerol gradient. The isolated nucleocapsids were then analyzed on SDS polyacrylamide gels to determine which of the four supernatant proteins bind to nucleocapsids.

the interrelationship of viral and host functions. Since transcription in VSV involves three viral proteins, and may involve RNA modification as well as transcription, it is of great interest to examine the effect of mutations in the proteins involved. Furthermore, there is accumulating evidence that protein(s) involved in transcription may also be involved in replication and the study of mutants can help elucidate if and how such protein(s) participate in both processes.

Various mutants have been studied in vitro to determine if a transcriptase defect could explain their in vivo properties. Attention has primarily focused on the *ts* mutants of VSV Indiana, but other mutants, e.g., host range mutants of VSV New Jersey,[49] a rifampin sensitive mutant of VSV New Jersey,[50] host range mutants of VSV Indiana,[51] and a plaque mutant of VSV Indiana have also been examined in vitro.[52] In the case of the last two types of mutant no obvious defect was discernible in vitro which would explain the in vivo phenotype but in the other groups mentioned at least some members had an altered transcriptase activity in vitro which appeared to be related to the in vivo defect.

Ts mutants of VSV Indiana fall into five well defined complementation groups with some evidence for a sixth group.[53-56] Mutants in Groups I and IV and certain members of Group II synthesize far less viral-specific RNA in vivo at the nonpermissive temperature than does wild type (*wt*) VSV.[57-59] Mutants with a defect in transcription are likely to belong to one, or more, of these groups. It has not been as easy to assign proteins to these groups as might initially have been expected.

In vitro transcription by certain Group I mutants is more temperature sensitive than that by *wt* VSV; some group IV mutants show a similar lesion, although this is by no means a general property of Group IV mutants.[60-64] Is the thermosensitivity in such mutants attributable to a mutation in the nucleoprotein template (i.e., the N protein) or in L or NS proteins? To investigate this, purified virions of Group I mutants were separated into supernatant (containing L plus NS proteins) and pellet (containing N protein and RNA) fractions. When homologous supernatant and pellet fractions were recombined the heat sensitivity of transcription was similar to that of unfractionated virions,[32,64,65] although in certain mutants the supernatant activity was too labile to survive the fractionation procedure.[65] In heterologous reconstitution experiments, *ts* pellet fractions combined with *wt* supernatant fractions showed a thermosensitivity typical of *wt* virus. However, the temperature sensitivity of a reaction containing *ts* supernatant plus *wt* pellet was similar to that of the unfractionated *ts* mutant,[32,64,65] indicating that the defect lay in one of the proteins present in the supernatant fraction — presumably L or NS. Two wild-type revertants of a Group I mutant behaved like *wt* virus in fractionation-reconstitution experiments.[65] Thus it seems that the thermosensitivity is correlated with the *ts* mutation.

To further locate the *ts* defect, L or NS proteins were fractionated from group I *ts* mutants and combined with NS or L proteins from *wt* virus in the presence of *wt* template.[32] Transcription by *ts* NS + *wt* L + *wt* template showed the same temperature sensitivity as *wt* NS + *wt* L + *wt* template, suggesting that the NS protein in Group I mutants is not temperature-sensitive. The activity of the L protein from the mutants was too unstable for the reciprocal experiment to be done, so "in vitro complementation" was used. Virions were dissociated by treatment with high-salt in the presence of detergent, then *wt* L or *wt* NS was added. The salt concentration was then lowered and transcription activity assayed. This procedure was used to ensure the added proteins did not have to compete with bound proteins for template. The results indicated that addition of either protein had little effect on transcription by *wt* virions or by a Group IV mutant which is temperature-sensitive for transcription in vitro (Table 2).[32] However, addition of *wt* L protein (but not *wt* NS protein) to Group I *ts* mutants resulted in a dramatic recovery of activity at 39°C compared to 31°C, suggesting that *wt* L protein can replace the defective protein in such mutants at the nonpermissive temperature. Hence it appears that Group I mutants have a defective L protein.

A Group IV mutant, *ts* W16B (IV) has been similarly fractionated into pellet and supernatant fractions and homologous and heterologous reconstitutions carried out as described for the Group I mutants. The defect is in the nucleoprotein fraction.[65] However, since in these experiments, this fraction contained appreciable amounts of NS as well as N, it is difficult to prove which protein is defective. Revertants of *ts* W16B show a wild type phenotype in such experiments so the pellet defect appears to be related to the in vivo properties of this mutant.

Temperature-dependent host range mutants of VSV New Jersey, which grow on BHK-21 cells at both 31°C and 39°C but grow on chick embryo fibroblasts (CEF) at 31°C only have been studied in vitro.[49] The transcriptase activity of certain of these mutants is more heat-sensitive than that of wild-type virus and the nature of the lesion in one such mutant (*td* CE 3) was examined.[66] When *wt* supernatant fractions containing predominantly L protein were combined with nucleoprotein fractions, containing mainly N and NS proteins, from *wt*, *td* CE 3, or a wild type revertant of *td* CE 3, the

TABLE 2

Temperature-sensitivity of the Transcriptase Activity of Solubilized, Unfractionated *wt* and Mutant VS Virus when Supplemented with *wt* L or *wt* NS Protein Fractions[a]

	% [^{3}H]UMP incorporated at restrictive temperature[b]					
	Controls[c]		Added *wt* L[d]		Added *wt* NS[e]	
	No additional:		0.05	0.10	0.125	0.250
Virus	*wt* L	*wt* NS	m*l*	m*l*	m*l*	m*l*
wt	21.5	21.9	21.2	20.4	27.6	24.0
ts G13 (I)[f]	2.2	7.3	17.6	26.1	−3.6	1.0
ts G16 (I)	−0.7	1.8	15.0	26.3	1.1	−0.2
ts G114 (I)	3.2	6.0	20.2	33.0	3.7	1.3
ts W16B (IV)	9.6	5.6	4.6	9.5	6.1	10.0

[a] Virus was solubilized with Triton-high salt but not fractionated. Phosphocellulose- and DEAE-fractionated *wt* L or *wt* NS protein (or the appropriate buffer) was then added, followed by buffer containing no salt and transcriptase assay mix containing nucleoside triphosphates and magnesium. [^{3}H]UMP incorporation into acid-insoluble material at 31 or 39°C was then monitored.

[b] (Counts/min of [^{3}H]UMP incorporated in 120 min at 39°C) × 100/ (counts/min of [^{3}H]UMP incorporated in 120 min at 31°C).

[c] As *wt* L and *wt* NS proteins are in buffers of different salt concentrations, the procedure for setting up the assay varies slightly according to whether *wt* L or *wt* NS is to be added to the solubilized virus. Two separate control assays were therefore set up for each experiment.

[d] Milliliter of *wt* L added per 1.7 m*l* of transcriptase assay mix.

[e] Milliliter of *wt* NS added per 1.7 m*l* of transcriptase assay mix.

[f] The roman numerals in parentheses refer to the complementation group to which each mutant belongs.

From Hunt, D. M., Emerson, S. U., and Wagner, R. R., *J. Virol.*, 18, 596, 1976. With permission.

heat-sensitivity of all such combinations was phenotypically wild type. Therefore, it appears that the nucleocapsid is not defective. When an L protein fraction from *wt*, CE 3, or a revertant, was combined with a N and NS containing nucleocapsid, the *td* CE 3 L protein was the only one which showed a *ts* rather than a *wt* phenotype. Thus the in vivo properties of *td* CE 3 appear to be connected with an abnormal L protein.

Moreau and co-workers have studied a rifampin-sensitive mutant of VSV Indiana and compared it to a rifampin-resistant revertant and wild type VSV (which is also rifampin-resistant).[50,67,68] The transcriptase activity present in virions of the rifampin-sensitive mutant is sensitive to rifampin in vitro, unlike the enzyme present in the revertant or *wt* VSV. This correlates with abnormalities in transcription in vivo by the mutant, suggesting an alteration in the transcription system.

VIII. CORRELATION OF IN VITRO TRANSCRIPTION DATA WITH IN VIVO DATA

Unfortunately, although the in vitro systems currently in use can be informative, they are not ideal model systems for in vivo reactions and care must be taken in interpreting and assessing the significance of the data. The temperature optimum for transcription by wild type VSV in vitro (28 to 31°C) is lower than that for transcription by input nucleocapsids in vivo (34 to 40°C).[8,69,70] Furthermore, although transcription is complete in cells productively infected at 37°C or 39°C, in in vitro experiments at 37°C only 25 to 50% of the viral genome was transcribed.[71] It would be of interest to repeat these experiments in view of developments in assay systems made to favour stabilization of transcription at higher temperatures.[63,64]

As already discussed, in vitro experiments suggest that group I VSV-Indiana mutants have a defective L protein. This correlates well with in vivo results in which Group I mutants have been found to make little or no viral specific RNA at the nonpermissive temperature.[57-59,62,70,72,73] Flamand and Bishop,[70,73] using a highly sensitive hybridization technique, showed that certain Group I mutants carry out some transcription at the nonpermissive temperature although secondary transcription (amplification) did not occur. However, no transcription at the nonpermissive temperature is detectable when at least one mutant (*ts* G114 [I]) is assayed by this method.[31] This suggests that in this mutant the transcriptase is nonfunctional in vivo and is therefore likely to be the moiety affected in all Group I mutants. This correlates with the in vitro finding of a defective L. Moreover, Knipe et al. found that the L protein synthesized in vivo by Group I mutants was unstable in vivo.[74]

One would not expect all mutations in the transcription system to result in a simple transcription negative phenotype. The assays used do not take into account all the subtleties of the system. It is always possible that a defect in, for instance, the L protein could result in the synthesis of detectable amounts of RNA at the nonpermissive temperature but that this RNA does not function normally. It would be of interest to survey Group I mutants which do make RNA and to try to analyze the defect in these. This might give us some clue as to the involvement of L in ancillary processes of transcription such as cleavage and capping. In some mutants it might be the temperature at which the L protein is made and folded which is the more important than the temperature at which its function is investigated.[70]

A further complication with the transcription-defective mutants is that Group I might represent mutants in more than one protein, since if two transcription-negative mutants with a defect in different proteins are to complement each other, protein exchange between input virions will be necessary, and this may not be efficient.[72]

Correlation of the in vivo and in vitro evidence for group IV mutants is more difficult. *Ts* W16B can synthesize RNA in vivo at the nonpermissive temperature and it seems that replication is blocked,[59] but in vitro experiments suggest that a transcription defect is connected with the *ts* phenotype.[65] It might be that a protein present in the transcription complex is also functional in replication. A mutation in the latter function may or may not affect transcription. It could be that, in vitro, when the transcriptase activity seems generally more labile, a lesion in this protein which does not affect in vivo transcription significantly becomes accentuated, affecting in vitro activity. Candidates for such a protein are N or NS. There is conflicting data as to whether Group IV represents N or NS mutants.[74,75]

Various other experiments suggest some interconnection between transcription and replication. For instance, although the major defect in Group IV mutants appears to be a block in replication,[59,72,76] certain members show abnormalities in in vivo transcription at the nonpermissive temperatures as well.[62,76,77] Similarly, with certain Group I mutants there is evidence for defective replication as well as transcription,[31,62] although some mutants do retain the ability to make 40S progeny RNA at the nonpermissive temperature.[76,78]

Addition of cycloheximide or other inhibitors of protein synthesis (which also results in a block in synthesis of progeny 40S RNA) has interesting results. In a large plaque variant of VSV (but not a small plaque mutant) this results in increased transcription.[52]

In some Group I mutants, if cycloheximide is added at the time a productively infected culture is shifted up to the nonpermissive temperature, instead of the almost complete inhibition of mRNA synthesis seen in the absence of drug, considerable transcription occurs.[78] Similarly, transcription at the nonpermissive temperature is stimulated in certain Group IV mutants if cycloheximide is added at the time of shift-up.[77] It has been suggested that these results could reflect some balance between replication and transcription; however, a completely satisfying explanation for these tantalizing results is not yet forthcoming. The diversity of results with the *ts* transcription-replication defective mutants is not too surprising in view of the apparently multifunctional nature of the transcriptase; since it is likely that transcriptase components are also involved in the synthesis of the 40S plus-strand RNA-template for replication,[79] control by a variety of factors is possible.

IX. INVOLVEMENT OF HOST FACTORS IN TRANSCRIPTION

There is some evidence to suggest that host factor(s) might be involved in transcription in vivo. It is not known what these factors are, if indeed they exist. There have been reports that addition of various cell extracts results in increased viral transcription in vitro.[21,80,81] However, such factors are not necessary for synthesis of full-length, active mRNA in an in vitro system,[44] and it is not clear whether they merely provide, for example, RNAse inhibitors and a favourable milieu for the enzyme or whether they provide some more concrete function such as a host protein which interacts with the transcriptase complex. Since transcription by input nucleocapsids occurs under conditions in which host RNA and protein synthesis is inhibited,[73] if any host function is necessary in vivo it must be preexisting at the time of infection and not virus-induced.

In the temperature-dependent host range mutant *td* CE 3 and its wild type revertant studied by Szilágyi and co-workers,[49,66] it appeared that an alteration in the L protein was directly related to the loss in the ability of the virus to grow on CEF cells at 39°C. Since the virus could still grow on BHK-21 cells at 39°C it appears that BHK-21 cells either provide a factor(s) which enables the L protein to function at 39°C in vivo, or lack a factor(s) which leads to its de-stabilization and which is present in CEF cells. It is not known if this factor is the general cell environment or has a more specific nature.

X. SUMMARY

Most, if not all, rhabdoviruses carry an RNA-dependent RNA polymerase or trnscriptase in the virion. This enzyme is responsible for transcribing virion RNA into mRNA and is probably required to initiate viral reproduction. Three viral proteins (L, NS, and N) compose the ribonucleocapsid of vesicular stomatitis virus and all are obligatory for transcription. Experiments with mutants suggest that transcription and replication may utilize some or all of the same viral proteins; in addition, it is possible that host cell proteins are involved. In spite of the impressive amount of data accumulated, we still do not know the stoichiometry of the active transcription complex or the specific function of each of the required viral proteins.

ACKNOWLEDGMENTS

This work was supported by Public Health Service grants AI-11722 and AI-05292 from the National Institute of Allergy and Infectious Diseases. S.U.E. is the recipient of Public Health Service Research Career Development Award AI-00013 from the National Institute of Allergy and Infectious Disease. We wish to thank L. Kingsford, H. Belle-Isle, and A. Carroll for helpful discussions and criticisms.

REFERENCES

1. **Kates, J. R. and McAuslan, B. R.**, Poxvirus DNA-dependent RNA polymerase, *Proc. Nat. Acad. Sci. U.S.A.*, 58, 134, 1967.
2. **Munyon, W., Paoletti, E., and Grace, J. T.**, RNA polymerase activity in purified infectious vaccinia virus, *Proc. Nat. Acad. Sci. U.S.A.*, 58, 2280, 1967.
3. **Shatkin, A. J. and Sipe, J. D.**, RNA polymerase activity in purified reoviruses, *Proc. Nat. Acad. Sci. U.S.ja.*, 61, 1462, 1968.
4. **Borsa, J. and Graham, A. F.**, Reovirus: RNA polymerase activity in purified virions, *Biochem. Biophys. Res. Commun.*, 33, 895, 1968.
5. **Lewandowski, L. J., Kalmakoff, J., and Tanada, Y.**, Characterization of a ribonucleic acid polymerase activity associated with purified cytoplasmic polyhedrosis virus of the silkworm *Bombyx mori*, *J. Virol.*, 4, 857, 1969.
6. **Baltimore, D., Huang, A. S., and Stampfer, M.**, Ribonucleic acid synthesis of vesicular stomatitis virus, II. An RNA polymerase in the virion, *Proc. Nat. Acad. Sci. U.S.A.*, 66, 572, 1970.
7. **Banerjee, A. K.**, Volume II.
8. **Aaslestad, H. G., Clark, H. F., Bishop, D. H. L., and Koprowski, H.**, Comparison of the ribonucleic acid polymerases of two rhabdoviruses, Kern Canyon virus and vesicular stomatitis virus, *J. Virol.*, 7, 726, 1971.
9. **Chang, S. H., Hefti, E., Obijeski, J. F., and Bishop, D. H. L.**, RNA transcription by the virion polymerases of five rhabdoviruses, *J. Virol.*, 13, 652, 1974.
10. **Roy, P., Clark, H. F., Madore H. P., and Bishop, D. H. L.**, RNA polymerase associated with virions of pike fry rhabdoviruses, *J. Virol.*, 15, 338, 1975.
11. **McAllister, P. E. and Wagner, R. R.**, Virion polymerases of two salmonid rhabdoviruses, *J. Virol.*, 22, 839, 1977.
12. **Francki, R. I. B. and Randles, J. W.**, RNA-dependent RNA polymerase associated with particles of lettuce necrotic yellows virus, *Virology*, 47, 270, 1972.
13. **Kawai, A.**, Transcriptase activity associated with rabies virions, *J. Virol.*, 24, 826, 1977.
14. **Follett, E. A. C., Pringle, C. R., Wunner, W. H., and Skehel, J. J.**, Virus replication in enucleate cells: vesicular stomatitis virus and influenza virus, *J. Virol.*, 13, 394, 1974.
15. **Rose, J.**, Volume II.
16. **Pringle, C. R. and Szilágyi, J.**, Volume II.
17. **Bishop, D. H. L. and Roy, P.**, Dissociation of vesicular stomatitis virus and relation of the virion proteins to the viral transcriptase, *J. Virol.*, 10, 234, 1972.
18. **Emerson, S. U. and Wagner, R. R.**, Dissociation and reconstitution of the transcriptase and template activities of vesicular stomatitis B and T virions, *J. Virol.*, 10, 297, 1972.
19. **Szilágyi, J. F. and Uryvayev, L.**, Isolation of an infectious ribonucleoprotein from vesicular stomatitis virus containing an active RNA transcriptase, *J. Virol.*, 11, 279, 1973.
20. **Abraham, G. and Banerjee, A. K.**, The nature of the RNA products synthesized *in vitro* by subviral components of vesicular stomatitis virus, *Virology*, 71, 230, 1976.
21. **Preston, C. M. and Szilágyi, J. F.**, Cell-free translation of RNA synthesized *in vitro* by a transcribing nucleoprotein complex prepared from purified vesicular stomatitis virus, *J. Virol.*, 21, 1002, 1971.
22. **Galet, H. and Prevec, L.**, Polyadenylate synthesis by extracts from L-cells infected with vesicular stomatitis virus, *Nature (London) New Biol.*, 243, 200, 1973.
23. **Galet, H., Hallet, D., and Prevec, L.**, Analysis of *in vitro* transcription products of intracellular vesicular stomatitis virus RNA polymerase, *J. Virol.*, 19, 467, 1976.
24. **Murphy, M. F. and Lazzarini, R. A.**, Synthesis of viral mRNA and polyadenylate by a ribonucleoprotein complex from extracts of vesicular stomatitis virus infected cells, *Cell*, 3, 78, 1974.
25. **Bishop, D. H. L., Emerson, S. U., and Flamand, A.**, Reconstitution of infectivity and transcriptase activity of homologous and heterologous viruses: vesicular stomatitis (Indiana serotype), Chandipura, vesicular stomatitis (New Jersey serotype) and Cocal viruses, *J. Virol.*, 14, 139, 1974.
26. **Emerson, S. U. and Yu, Y. H.**, Both NS and L proteins are required for *in vitro* RNA synthesis by vesicular stomatitis virus, *J. Virol.*, 15, 1348, 1975.
27. **Emerson, S. U. and Wagner, R. R.**, L protein requirement for *in vitro* RNA synthesis by vesicular stomatitis virus, *J. Virol.*, 12, 1325, 1973.
28. **Naito, S. and Ishihama, A.**, Function and structure of RNA polymerase from vesicular stomatitis virus, *J. Biol. Chem.*, 251, 4307, 1976.
29. **Bishop, D. H. L. and Roy, P.**, Kinetics of RNA synthesis by vesicular stomatitis virus particles, *J. Mol. Biol.*, 57, 513, 1971.
30. **Emerson, S. U.**, unpublished data, 1977.

31. **Repik, P., Flamand, A., and Bishop, D. H. L.,** Synthesis of RNA by mutants of vesicular stomatitis virus (Indiana serotype) and the ability of the wild-type VSV New Jersey to complement the VSV-Indiana *ts*G I-114 transcription defect, *J. Virol.*, 20, 157, 1976.
32. **Hunt, M. D., Emerson, S. U., and Wagner, R. R.,** RNA temperature-sensitive mutants of vesicular stomatitis virus: L-protein thermosensitivity accounts for transcriptase restriction of Group I mutants, *J. Virol.*, 18, 596, 1976.
33. **Wagner, R. R.,** Reproduction of rhabdoviruses, in *Comprehensive Virology*, Vol. 4, Fraenkel-Conrat, H. and Wagner, R. R., Eds., Plenum Press, 1975.
34. **Wagner, R. R., Schnaitman, T. C., and Snyder, R. M.,** Structural proteins of vesicular stomatitis viruses, *J. Virol.*, 3, 395, 1969.
35. **Kang, C. Y. and Prevec, L.,** Proteins of vesicular stomatitis virus. I. Polyacrylamide gel analysis of viral antigens, *J. Virol.*, 3, 304, 1969.
36. **Stamminger, G. and Lazzarini, R. A.,** Analysis of the RNA of defective VSV particles, *Cell*, 3, 85, 1974.
37. **Schnitzlein, W. M. and Reichmann, M. E.,** The size and the cistronic origin of defective vesicular stomatitis virus particle RNAs in relation to homotypic and heterotypic interference, *J. Mol. Biol.*, 101, 307, 1976.
38. **Colonno, R. J., Lazzarini, R. A., Keene, J. D., and Banerjee, A. K.,** *In vitro* synthesis of messenger RNA by a defective interfering particle of vesicular stomatitis virus, *Proc. Nat. Acad. Sci. U.S.A.*, 74, 1884, 1977.
39. **Emerson, S. U., Dierks, P. M., and Parsons, J. T.,** RNA Synthesis by a defective particle of vesicular stomatitis virus, in *Negative Strand Viruses*, Mahy, B. W. J. and Barry, R. D., Eds., Academic Press, London, 1978, in press.
40. **Emerson, S. U., Dierks, P. M., and Parsons, J. T.,** *In vitro* synthesis of a unique RNA species by a T particle of vesicular stomatitis virus, *J. Virol.*, 23, 708, 1977.
41. **Stampfer, M. and Baltimore, D.,** Identification of the vesicular stomatitis virus large protein as a unique viral protein, *J. Virol.*, 11, 520, 1973.
42. **Travers, A. A. and Burgess, R. R.,** Cyclic reuse of the RNA polymerase factor, *Nature (London)*, 222, 537, 1969.
43. **Sokol, F. and Clark, H. F.,** Phosphoproteins, structural components of rhabdoviruses, *Virology*, 52, 246, 1973.
44. **Both, G. W., Moyer, S. A., and Banerjee, A. K.,** Translation and identification of the mRNA species synthesized *in vitro* by virion-associated RNA polymerase of vesicular stomatitis virus, *Proc. Nat. Acad. Sci. U.S.A.*, 72, 274, 1975.
45. **Rhodes, D. P., Abraham, G., Colonno, R. J., Jelinek, W., and Banerjee, A. K.,** Characteristics of vesicular stomatitis virus mRNA species synthesized *in vitro*, *J. Virol.*, 21, 1105, 1977.
46. **Freeman, G. J., Rose, J. K., Clinton, G. M., and Huang, A. S.,** RNA synthesis of vesicular stomatitis virus. VII. Complete separation of the mRNAs of vesicular stomatitis virus by duplex formation, *J. Virol.*, 21, 1094, 1977.
47. **Imblum, R. L. and Wagner, R. R.,** Inhibition of viral transcriptase by immunoglobulin directed against the nucleocapsid NS protein of vesicular stomatitis virus, *J. Virol.*, 15, 1357, 1975.
48. **Mellon, M. G. and Emerson, S. U.,** The binding of L and NS proteins to the ribonucleocapsid of vesicular stomatitis virus, in *Negative Strand Viruses*, Mahy, B. W. J. and Barry, R. D., Eds., Academic Press, London, 1978, in press.
49. **Szilágyi, J. F. and Pringle, C. R.,** Virion transcriptase activity differences in host range mutants of vesicular stomatitis virus, *J. Virol.*, 16, 927, 1975.
50. **Moreau, M. C.,** Inhibition of a vesicular stomatitis virus mutant by rifampin, *J. Virol.*, 14, 517, 1974.
51. **Simpson, R. W. and Obijeski, J. F.,** Conditional lethal mutants of vesicular stomatitis virus. I. Phenotypic characterization of single and double mutants exhibiting host restriction and temperature-sensitivity, *Virology*, 57, 357, 1974.
52. **Wertz, G. W. and Levine, M.,** RNA synthesis by vesicular stomatitis virus and a small plaque mutant: effects of cycloheximide, *J. Virol.*, 12, 253, 1973.
53. **Pringle, C. R.,** Genetic characteristics of conditional lethal mutants of vesicular stomatitis virus induced by 5-fluoro-uracil; 5-azacytidine, and ethyl methane sulfonate, *J. Virol.*, 5, 559, 1970.
54. **Flamand, A. and Pringle, C. R.,** The homologies of spontaneous and induced temperature-sensitive mutants of vesicular stomatitis virus isolated in chick embryo and BHK-21 cells, *J. Gen. Virol.*, 11, 81, 1971.
55. **Cormack, D. V., Holloway, A. F., and Pringle, C. R.,** Temperature-sensitive mutants of vesicular stomatitis virus: homology and nomenclature, *J. Gen. Virol.*, 19, 295, 1973.
56. **Rettenmier, C. W., Dumont, R., and Baltimore, D.,** Screening procedure for complementation-dependent mutants of vesicular stomatitis virus, *J. Virol.*, 15, 41, 1975.

57. **Pringle, C. R. and Duncan, I. B.**, Preliminary physiological characterization of temperature-sensitive mutants of vesicular stomatitis virus, *J. Virol.*, 8, 56, 1971.
58. **Lafay, F.**, Etudes des mutants thermosensibles du virus de la stomatite vésiculaire (VSV). Classification de quelques mutants d'après des criteres de functionnement, *C. R. Acad. Sci. Paris Ser. D.*, 268, 2385, 1969.
59. **Wong, P. K. Y., Holloway, A. F., and Cormack, D. V.**, Characterization of three complementation groups of vesicular stomatitis virus, *Virology*, 50, 829, 1972.
60. **Cormack, D. V., Holloway, A. F., Wong, P. K. Y., and Cairns, J. E.**, Temperature-sensitive mutants of vesicular stomatitis virus. II. Evidence of defective polymerase, *Virology*, 45, 824, 1971.
61. **Cairns, J. E., Holloway, A. F., and Cormack, D. V.**, Temperature-sensitive mutants of vesicular stomatitis virus: *in vitro* studies of virion-associated polymerase, *J. Virol.*, 10, 1130, 1972.
62. **Printz-Ané, C., Combard, A., and Martinet, C.**, Study of the transcription and replication of vesicular stomatitis virus by using temperature-sensitive mutants, *J. Virol.*, 10, 889, 1972.
63. **Szilágyi, J. F. and Pringle, C. R.**, Effect of temperature-sensitive mutations on the virion-associated RNA transcriptase of vesicular stomatitis virus, *J. Mol. Biol.*, 71, 281, 1972.
64. **Hunt, D. M. and Wagner, R. R.**, Location of the transcription defect in group I temperature-sensitive mutants of vesicular stomatitis virus, *J. Virol.*, 13, 28, 1974.
65. **Ngan, J. S. C., Holloway, A. F., and Cormack, D. V.**, Temperature-sensitive mutants of vesicular stomatitis virus: comparison of the *in vitro* RNA polymerase defects of group I and group IV mutants, *J. Virol.*, 14, 765, 1974.
66. **Szilágyi, J. F., Pringle, C. R., and MacPherson, T. M.**, Temperature-dependent host range mutation in vesicular stomatitis virus affecting polypeptide L, *J. Virol.*, 22, 381, 1977.
67. **Moreau, M. C., Sanzey, B., and Chany, C.**, *In vitro* transcription of a vesicular stomatitis virus mutant, in *In vitro* Transcription and Translation of Viral Genomes, Haenni, A. L. and Beaud, G., Eds., INSERM, Paris, 1975, 103.
68. **Moreau, M. C. and Sanzey, B.**, Rifampin-susceptible mutant of vesicular stomatitis virus: protein and RNA synthesis. *J. Virol.*, 21, 41, 1977.
69. **Huang, A. S. and Manders, E. K.**, Ribonucleic acid synthesis of vesicular stomatitis virus. IV. Transcription by standard virus in the presence of defective interfering particles, *J. Virol.*, 9, 909, 1972.
70. **Flamand, A. and Bishop, D. H. L.**, Primary *in vivo* transcription of vesicular stomatitis virus and temperature-sensitive mutants of five vesicular stomatitis virus complementation groups, *J. Virol.*, 12, 1238, 1973.
71. **Bishop, D. H. L. and Roy, P.**, Properties of the product synthesized by vesicular stomatitis virus particles, *J. Mol. Biol.*, 58, 799, 1971.
72. **Unger, J. T. and Reichmann, M. E.**, RNA synthesis in temperature-sensitive mutants of vesicular stomatitis virus, *J. Virol.*, 12, 570, 1973.
73. **Flamand, A. and Bishop, D. H. L.**, *In vivo* synthesis of RNA by vesicular stomatitis virus and its mutants, *J. Mol. Biol.*, 87, 31, 1974.
74. **Knipe, D., Lodish, H. F., and Baltimore, D.**, Analysis of the defects of temperature-sensitive mutants of vesicular stomatitis virus: intracellular degradation of specific viral proteins, *J. Virol.*, 21, 1140, 1977.
75. **Combard, A., Printz-Ane, C., Martinet, C., and Printz, P.**, Temperature-sensitive defect of vesicular stomatitis virus in complementation group II, *J. Virol.*, 21, 913, 1977.
76. **Perlman, S. M. and Huang, A. S.**, Virus-specific RNA specified by the group I and group IV temperature-sensitive mutants of vesicular stomatitis virus, *Intervirology*, 2, 312, 1973/1974.
77. **Combard, A., Martinet, C., Printz-Ané, C., Friedman, A., and Printz, P.**, Transcription and replication of vesicular stomatitis virus: effects of temperature-sensitive mutations in complementation group IV, *J. Virol.*, 13, 922, 1974.
78. **Perlman, S. M. and Huang, A. S.**, RNA synthesis of vesicular stomatitis virus. V. Interactions between transcription and replication, *J. Virol.*, 12, 1395, 1973.
79. **Soria, M., Little, S. P., and Huang, A. S.**, Characterization of vesicular stomatitis virus nucleocapsids. I. Complementary 40S RNA molecules in nucleocapsids, *Virology*, 61, 270, 1974.
80. **Ball, L. A. and White, C. N.**, Order of transcription of genes of vesicular stomatitis virus, *Proc. Nat. Acad. Sci. U.S.A.*, 73, 442, 1976.
81. **Rose, J. K., Lodish, H. F., and Brock, M.**, Giant heterogeneous polyadenylic acid on vesicular stomatitis virus mRNA synthesized *in vitro* in the presence of S-adenosylhomocysteine, *J. Virol.*, 21, 683, 1977.

INDEX

G

H

I

J

M

N

Q

R

S

T

U

V

W

Y

For Product Safety Concerns and Information please contact our EU
representative GPSR@taylorandfrancis.com
Taylor & Francis Verlag GmbH, Kaufingerstraße 24, 80331 München, Germany

www.ingramcontent.com/pod-product-compliance
Ingram Content Group UK Ltd.
Pitfield, Milton Keynes, MK11 3LW, UK
UKHW052052230926
475702UK00025B/1141
* 9 7 8 1 3 1 5 8 9 7 2 8 8 *